FOREWORD

The OECD/NEA Nuclear Science Committee set up a Working Party on Physics of Plutonium Recycling in June 1992 to deal with the status and trends of physics issues related to plutonium recycling with respect to both the back end of the fuel cycle and the optimal utilisation of plutonium. For completeness, issues related to the use of the uranium coming from recycling are also addressed.

The Working Party met three times and the results of the studies carried out have been consolidated in the series of reports "Physics of Plutonium Recycling".

The series covers the following aspects:

- Volume I *Issues and Perspectives*;
- Volume II *Plutonium Recycling in Pressurized-Water Reactors*;
- Volume III *Void Reactivity Effect in Pressurized-Water Reactors*;
- Volume IV *Fast Plutonium-Burner Reactors: Beginning of Life*;
- Volume V *Plutonium Recycling in Fast Reactors*; and,
- Volume VI *Multiple Recycling in Advanced Pressurized-Water Reactors.*

The present volume is the third in the series and describes the specific benchmark study concerned with the void reactivity effect in MOX-fuelled pressurized-water reactors.

CONTENTS

SUMMARY

The report gives an overview of the solutions of the Void Reactivity Effect Benchmark conducted by OECD/NEA, submitted by 18 participants from 12 institutes in 8 countries. The results are briefly discussed and presented in plots and tables. The infinite lattice calculations with uranium and different enriched MOX fuels give a non-negligible spread of the results, but a clear tendency of the void effect, which is positive for the high content of MOX in the fuel and becomes negative with decreasing plutonium density. A similar tendency is derived from the results of two-dimensional calculations for an uranium assembly with a central moderated or voided region fuelled with MOX pins of different plutonium content. The results show a larger spread than for the infinite lattice case, caused by small differences in the eigenvalues of moderated and voided cases.

CONTRIBUTORS

AUTHORS	*D. Lutz*	IKE	Germany
	W. Bernnat	IKE	Germany
	K. Hesketh	BNFL	U.K.
	E. Sartori	OECD/NEA	
PROBLEM SPECIFICATION	*Th. Maldague*	Belgonucléaire	Belgium
	G. Minsart	SCK-CEN	Belgium
	P. D'hondt	SCK-CEN	Belgium
DATA COMPILATION AND ANALYSIS	*D. Lutz*	IKE	Germany
	A. & W. Bernnat	IKE	Germany
	M. Mattes	IKE	Germany
SUPPLEMENTARY ANALYSIS	*S. Cathalau*	CEA	France
	M. Soldevila	CEA	France
	S. Ralphs	CEA	France
	A. Maghnouj	CEA	France
	G. Rimpault	CEA	France
	Ph. J. Finck	CEA	France
	H. Takano	JAERI	Japan
	T. Mori	JAERI	Japan
	H. Akie	JAERI	Japan
TEXT PROCESSING AND OUTLAY	*P. Jewkes*	OECD/NEA	

BENCHMARK PARTICIPANTS

Th. Maldague	Belgonucléaire	Belgium
G. Minsart	SCK-CEN	Belgium
S. Cathalau	CEA	France
A. Maghnouj	CEA	France
S. Rahlfs	CEA	France
G. Rimpault	CEA	France
M. Soldevila	CEA	France
M. Aigle	Framatome	France
W. Bernnat	IKE	Germany
S. Käfer	IKE	Germany
D. Lutz	IKE	Germany
M. Mattes	IKE	Germany
W. Hetzelt	Siemens	Germany
G. Schlosser	Siemens	Germany
P. A. Landeyro	ENEA	Italy
K. Ishii	Hitachi Ltd.	Japan
H. Maruyama	Hitachi Ltd.	Japan
H. Akie	JAERI	Japan
K. Kaneko	JAERI	Japan
K. Okumura	JAERI	Japan
H. Takano	JAERI	Japan
Y. Uenohara	Toshiba Corp.	Japan
J. K. Aaldijk	ECN	Netherlands
W. E. Freudenreich	ECN	Netherlands
A. Hogenbirk	ECN	Netherlands
A. Tsibulia	IPPE	Russia
R. N. Blomquist	ANL	U.S.A.

Introduction

The use of MOX fuel in standard PWRs introduces inhomogeneities in the core, which cause difficulties with the calculation of power distribution around the interfaces between UO_2 and MOX assemblies. This problem is well known [1,2]. A correct treatment needs improvements of the calculational models. This report deals with another problem connected with the use of plutonium fuel discussed for future modifications of the fuel cycle, namely the behaviour of the void coefficient for different MOX fuel types. The objective of the proposed benchmark is to check computer codes and nuclear data rather than to represent a realistic situation. The partially voided arrangement of assemblies is difficult to calculate and represents a challenge to designers of codes for assembly calculations.

The benchmark can be calculated by continuous energy Monte Carlo codes to get a best estimate solution, which essentially depends only on the data base used.

Benchmark specifications

A complete specification of the void reactivity effect benchmark problem can be found in Appendix A. The void reactivity effect benchmark specifies a supercell configuration of a 30×30 array of PWR fuel cells, with reflective boundary conditions. The central 10×10 region consists of either UO_2 or MOX rods (with three different plutonium contents) the configurations of which alternate between full moderation and complete voidage of the moderator. In every case the outer part of the macrocell is assumed to be fully moderated. The configurations with UO_2 rods (3.35 w/o U-235), MOX rods of high enrichment (14.4 w/o total Pu), MOX rods of medium enrichment (9.7 w/o total Pu) and MOX rods of low enrichment (5.4 w/o total Pu) in the central 10×10 sub-assembly are designated UO_2, H-MOX, M-MOX and L-MOX respectively.

Participants, methods and data

Eighteen solutions were contributed for the benchmark by 12 institutions in 8 countries. A short description of codes and data bases used by participants is provided in the following and is summarised in Table 1. The abbreviation in brackets behind the name of each institution is the identifier used in tables and figures.

1. Argonne National Laboratory, (ANL), U.S.A.

Participant:	R. N. Blomquist
Code:	VIM (continuous Monte Carlo)
Data Library:	ENDF/B-V
Remarks:	Details are in Appendix B.1.

2. ***Belgonucléaire (BEN), Belgium***

Participant:	Th. Maldague
Code:	LWRWIMS
Data Library:	1986 WIMS

3. ***Commissariat à l'Energie Atomique (CEA3) and Framatome, France***

Participants:	M. Soldevila and M. Aigle
Code:	APOLLO-2
Data Library:	JEF-2.2, CEA-93
Remarks:	Details are in Appendix B.2

4. ***Commissariat à l'Energie Atomique (CEA4), France***

Participants:	G. Rimpault and S. Rahlfs
Code:	ECCO-5.2
Data Library:	JEF-2.2
Remarks:	Details are in Appendix B.3

5. ***Commissariat à l'Energie Atomique (CEA5), France***

Participants:	S. Cathalau and A. Maghnouj
Code:	APOLLO-1 (P_{ij}) + S_n-Code
Data Library:	CEA-86 (ENDF/B-V + JEF-1)
Remarks:	Details are in Appendix B.4

6. ***Centre d'Etude de l'Energie Nucléaire (CEN), Belgium***

Participant:	G. Minsart
Code:	DTF-4/DOT-3,5
Data Library:	MOL-BR2
Remarks:	Details are published in a distributed report GM/gm–34.B4214, 49 WPPR/93-28 [3]

7. ***ECN Nuclear Energy (ECN), Netherlands***

Participants:	W. E. Freudenreich, J. K. Aaldijk and A. Hogenbirk
Code:	MCNP-4.2
Data Library:	JEF-2.2 and SCALE
Remarks:	Details are in published in the report ECN-R–94-007 [4]

8. ***ENEA, Energy Department (ENEA), Italy***

Participant:	P. A. Landeyro
Code:	MCNP-4.2
Data Library:	JEF-1
Remarks:	Details are in Appendix B.5

9. Hitachi Ltd. (HIT), Japan

Participants:	K. Ishii and H. Maruyama
Code:	VMONT
Data Library:	JENDL-2, ENDF/B-IV

10. University of Stuttgart (IKE1), Germany

Participant:	D. Lutz
Code:	CGM, RSYST
Data Library:	JEF-1
Remarks:	Details are in Appendix B.6

11. University of Stuttgart (IKE2), Germany

Participants:	W. Bernnat, M. Mattes and S. Käfer
Code:	MCNP-4.2
Data Library:	JEF-2.2
Remarks:	Details are in Appendix B.7

12. Institute of Physics and Power Engineering (IPPE), Russia

Participant:	A. Tsibulia
Code:	WIMS/D-4
Data Library:	FOND-2

13. Japan Atomic Energy Research Institute (JAE1), Japan

Participants:	H. Takano, H. Akie and K. Kaneko
Code:	SRAC
Data Library:	JENDL-3.1
Remarks:	Details are in Appendix B.8

14. Japan Atomic Energy Research Institute (JAE2), Japan

Participants:	H. Takano, H. Akie and K. Kaneko
Code:	MVP
Data Library:	JENDL-3.1
Remarks:	Details are in Appendix B.8

15. Japan Atomic Energy Research Institute (JAE3), Japan

Participant:	K. Okumura
Code:	SRAC/PIK
Data Library:	JENDL-3.1
Remarks:	Details are in Appendix B.9

16. ***Japan Atomic Energy Research Institute (JAE4), Japan***

Participant:	K. Okumura
Code:	SRAC/MOSRA
Data Library:	JENDL-3.1
Remarks:	Details are in Appendix B.9

17. ***Siemens (SIE1), Germany***

Participants:	G. Schlosser and W. Hetzelt
Code:	CASMO-3
Data Library:	J70

18. ***Toshiba Corporation (TOS), Japan***

Participant:	Y. Uenohara
Code:	MCNP-4.2
Data Library:	JENDL-3.1
Remarks:	Details are in Appendix B.10

Contributions 12 and 17 have been withdrawn.

Table 2 summarises the information on the resonance treatment used by participants in their models. Details can be found in the appendices.

Results

Tables 3, 5, 7 and 9 collect together all the multiplication factors from the various contributors. Tables 3 and 5 give k-infinities for the central 10×10 sub-assembly considered in isolation as an infinite lattice, for the fully moderated and fully voided configurations respectively. These k-infinities are a useful indication of the underlying agreement of the nuclear data libraries and fine-group flux calculational methods, without complications arising from neutron leakage to and from the 10×10 sub-assembly. The bottom line in both of these tables gives the arithmetic averages of all the contributions and Tables 4 and 6 list the corresponding calculated/average (c/a) ratios. Figures 1 and 2 display the same information as Tables 3 and 5 in a graphical form, while Figure 5 shows the corresponding void effects, i.e., the difference in k-infinity between the fully moderated and the fully voided situations.

Referring to Table 3, the low k-infinities for the MOX configurations (when compared with the UO_2 configuration), reflect the increased thermal absorption in MOX assemblies. When comparing these with the corresponding voided k-infinities from Table 5, it appears clearly that the k-infinity for the UO_2 lattice decreases considerably in the voided situation, corresponding to a negative void reactivity effect. While the same is true for the L-MOX and M-MOX cases, though the effect is smaller in magnitude, the H-MOX case has a higher k-infinity in the voided case, corresponding to a positive void reactivity effect. The reason for this is not difficult to understand when considering that the H-MOX case, consisting of 14.4 w/o total plutonium with no moderator, resembles a fast reactor more than a water reactor, it is then no surprise that k-infinity has a high value. In the fast reactor-like spectrum, all the plutonium isotopes contribute to fissions and k-infinity increases almost linearly with total plutonium content. In contrast, in the fully moderated situation only the odd plutonium isotopes are fissionable and k-infinity increases much more slowly with increasing plutonium content due to the increasing contribution of absorption in the even isotopes.

Tables 7 and 9 list k-infinity values for the whole macrocell for the fully moderated and voided cases respectively and Tables 8 and 10 give the corresponding c/a values. Figures 3, 4 and 6 show k-infinities and void effects in a graphical form. Since the macrocell volume is largely composed of UO_2 pins and only one ninth of the pins is in the central sub-assembly, the overall k-infinity vary significantly less between the various configurations. A particular point to bear in mind is that only the central 10×10 sub-assembly is subject to voiding and that there is a significant source of thermal neutrons into the voided sub-assembly associated with the surrounding fully moderated UO_2 region.

Although the all-UO_2 macrocell shows a negative void reactivity effect, the averages for the three MOX cases all show positive void effect decreasing with the Pu content of the fuel and near zero for the L-MOX case. This should not be taken to imply that the contribution of MOX assemblies to void reactivity will be positive in the situation of a real reactor, except with high plutonium contents. There is no possibility that in a reactor the situation arises where one assembly is fully voided while its neighbours are fully moderated. In the artificial situation modelled in this benchmark, the role of thermal neutrons leaking from the fully moderated region into the voided region is crucial. In a fully moderated MOX assembly, the neutron diffusion length is such that the boundary effect from leakage, from the UO_2 to the MOX region extends at most to two or three rows of fuel pins. In the voided case the neutron diffusion lengths increase so that the transient neutron currents extend throughout the 10×10 sub-assembly. The effect of voiding is therefore to couple the central MOX rods more strongly to the UO_2 regions which are rich in thermal neutrons. The central MOX rods are thereby able to contribute more effectively to the overall multiplication factor in the voided condition.

For the L-MOX case the CEA participants performed additional qualitative analyses using perturbation theory. The results reported in Appendix C.1 are very helpful to understand the complex physics of this arrangement.

For each of the four configurations of the central 10×10 sub-assembly, Figures 7 to 14 show the fully moderated and voided fission density distributions for a horizontal traverse across the macrocell (designated AM) and for a diagonal traverse (designated AD), both starting from the centre with the value 1. The same information is also presented in Tables 15 to 30. The arithmetic average for each pin location has been calculated and is included in the tables and plotted in the figures. The fission density distribution plots provide an indication of the degree to which the various solutions are in agreement as regards spatial coupling across the UO_2/MOX interface. The voided configurations give a strong interface effect that is challenging for the various methods to calculate.

Finally, Figures 15 to 35 show the local flux spectra for the central pin, the corner pin and the pin at the mid-point of the boundary of the central 10×10 sub-assembly. The spectra have been normalised in the same way and plotted on the same scale to aid comparison. Some participants submitted 4 group spectra. They can be compared in the tabulated presentation of the participants as provided in the Appendices B.

Discussion

The multiplication factors of the infinite cell lattices show spreads of up to 4 % (Tables 3 to 6, Figures 1 and 2). If the outlying solutions are omitted the spread reduces to about 1 % for the moderated cells and 2 % for the unmoderated cells. The reason for the latter may be in some cases partly due to deficiencies in the spatial coupling methods or in approximations inherent with cylindricising the square cells. The fact that there is a significant spread for the moderated cells is indicative of differences in the nuclear data libraries. This is reinforced by the observation that even the Monte Carlo solutions have a significant spread; for the Monte Carlo codes there are no concerns over approximations in the pin cell homogenisation or in the transport method, the spreads are therefore directly attributable to the nuclear data. In every case results of participants applying the same data base, JEF-2.2, JENDL-3.1 or JEF-1, form close clusters compared with the overall spread and show common tendencies. Thus, for example, the JEF-2.2 evaluated dataset yields larger k-infinities than the average for the UO_2 fuel and lower k-infinities for MOX, whereas JENDL-3.1 shows the opposite tendency.

The multiplication factors for the whole macrocell, given in Tables 7 to 10 reflect the spreads seen on the infinite lattice results for moderated uranium cells, since uranium rods make up a fraction of 8/9 of the volume in the MOX configurations and these rods are fully moderated in all cases. The Monte Carlo methods are particularly valuable reference data points as they allow an accurate representation of the strong spatial dependence of the flux spectra in the vicinity of the voided/moderated boundary. Deterministic methods which rely on homogenising pin cells are particularly questionable at this boundary, especially if zero current is assumed at the pin cell boundaries in generating the homogenisation spectra.

The fission density curves in the Figures 7 to 14 show a satisfactory agreement for the fully moderated cases, but show apparently highly discrepant results for the voided cases as plotted. This is to some extent an artefact of normalising each curve at the centre pin, which though convenient for presenting the results, transfers the spreads entirely to the UO_2 region. In reality, the power distribution errors will be much smaller, as the spread really applies to the voided region. Since the fission rates are very small in the voided region, the absolute errors on fission rates are within 10% there. As might be expected, the largest deviations actually occur at pin number 5, which borders on the UO_2 region. It is in this location where transport and homogenisation errors are most significant with the deterministic methods. The Monte Carlo codes again provide a useful reference, since they are not subject to such errors and the spread of results from the Monte Carlo codes is indicative of that in the underlying nuclear data. New research work of JAERI reported in Appendix C.2, confirms this statement. Calculations of some of the benchmark cases applying the new database JENDL-3.2 gives results rather close to the values resulting from MCNP-4 calculations using JEF-2.2 data.

The flux spectra of the central MOX pin show marked depressions in the voided cases at the resonance energies of large U-238 and Pu resonances both due to the lack of scattering sources and to the shadowing by neighbouring rods. The thermal spectra of the central pins are also very highly depressed, but those of the corner and edge pins are much closer to those of the adjacent UO_2 rods due to the neutron current from the latter.

Conclusions

The objective of the void reactivity effect benchmark was to compare the performance of the nuclear data libraries and codes presently available on a problem involving the calculation of the void coefficient in a mixed UO_2/MOX macrocell. The spread of k-infinities for the fully moderated infinite lattice slightly exceeds 1%, which is considered excessive and should be improved. Solutions obtained with the same nuclear data libraries tend to be grouped together, which indicates that the difference in the libraries are largely responsible. The voided configurations show larger spreads; there are two likely causes: the use of approximate homogenisation and transport methods at the voided/moderated boundary, and differences in higher energy cross-sections, which have a greater importance in the voided situation.

Overall, however, there is substantial agreement as to the trend for the void reactivity to become more positive as the plutonium content of the MOX region increases. The infinite lattice results show that the inherent void reactivity of MOX assemblies becomes positive somewhere between 10 and 14 w/o total plutonium content, at least with the isotopic composition assumed here.

Similar results were obtained with the hybrid UO_2/MOX macrocell, which is dominated by the properties of the moderated UO_2 lattice which makes up the bulk of the macrocell. The same trend towards more positive void effects was found as the plutonium content of the MOX regions increases. Because of the artificial nature of the problem, with a fully voided MOX sub-assembly adjacent to a fully moderated UO_2 driver region, positive void reactivity effects were obtained even for the L-MOX case, a result which would not occur in the more realistic case with a more uniform void distribution.

Following a detailed examination of the results, it is recommended that the differences in actinide cross-sections of the JEF-2.2 and JENDL-3.1 nuclear data libraries should be evaluated closely in order to explain the differences seen in the Monte Carlo calculations.

Acknowledgements

Special thanks go to all the participants who were willing to devote time to this endeavour and to our colleagues at IKE who provided a great help in processing the results.

References

[1] J. C. Lefèbvre, J. Mondot and J. P.West,
"Benchmark Calculations of Power Distribution within Assemblies" (PDWA), NEACRP-L-336 (1991).

[2] C. Cavarec, J. F. Perron, D. Verwaerde and J. P. West,
"Benchmark Calculations of Power Distribution within Assemblies" HT-12/92093 A (1992).

[3] G. Minsart, "Void Coefficient Benchmark and Computational Exercise on Void Reactivity Effect in MOX Lattices", SCK-CEN, GM/gm-34.B4214, 49 WPPR/93-28, December 1993.

[4] W. E. Freudenreich, J. K. Aaldijk and A. Hogenbirk,
"Void Coefficient Benchmark", ECN-R-94-007 (1994).

Table 1 **Summary of participants**

INSTITUTE	COUNTRY	CODE	DATA BASE/LIBRARY	N° OF GROUPS	REMARKS
ANL	U.S.A.	VIM	ENDF/B-V	infinite	Zircaloy
BEN	Belgium	LWRWIMS	1986 WIMS	69	
CEA 3	France	APOLLO 2	JEF-2.2 CEA 93	172/99	
CEA 4	France	ECCO52/ERANOS	JEF-2.2 CEA 93	1968/172	
CEA 5	France	APOLLO 1	ENDF/BV+ JEF1 CEA 86	99	
CEN	Belgium	DTF4/DOT 3,5	MOL-BR2	40	
ECN	Netherlands	MCNP 4.2	JEF-2.2 SCALE	172	
ENEA	Italy	MCNP 4.2	JEF-1	infinite	
Hitachi	Japan	VMONT	JENDL-2/ENDF/B-IV	190	
IKE 1	Germany	CGM/RSYST	JEF-1	224/60	
IKE 2	Germany	MCNP 4.2	JEF-2.2	infinite	
IPPE	Russia	WIMSD4	FOND-2 WIMS/ABBN		withdrawn
JAERI 1	Japan	SRAC	JENDL-3.1	107	
JAERI 2	Japan	MVP	JENDL-3.1	infinite	
JAERI 3	Japan	SRAC/PIK	JENDL-3.1	107	
JAERI 4	Japan	SRAC/MOSRA	JENDL-3.1	107	
Siemens	Germany	CASMO 3	J70	70	withdrawn
Toshiba	Japan	MCNP 4.2	JENDL-3.1	infinite	

Table 2 **Information about resonance treatment**

ANL Continuous-energy Monte Carlo, shielding of unresolved resonances,

BEN Self-shielded cross-sections of U-isotopes, Pu-239 and Pu-240 (1 eV resonance only),

CEA Self- and mutual shielding for the main U, Pu and Zr isotopes, local effects included,

CEN Self-shielded cross-section of all U and Pu isotopes,

ECN All actinides, fission products and Zr are self-shielded,

ENEA Continuous-energy Monte Carlo,

HIT Self-shielding for all actinides,

IKE 1 Self- and mutual shielding for the main U and Pu isotopes, by performing an ultrafine group cell calculation,

IKE 2 Continuous-energy Monte Carlo, no special treatment of unresolved resonances,

JAE Self- and mutual shielding for all U, Pu, Am isotopes by performing an ultrafine group cell calculation,

JAE2 Continuous-energy Monte Carlo,

TOS Continuous-energy Monte Carlo.

Table 3 : k-infinity, Pincell moderated

Contr.	Code	Data	Fuel Type			
			UO_2	H-MOX	M-MOX	L-MOX
ANL	VIM	ENDFB5	1.3651±.0010	1.2124±.0012	1.1671±.0011	1.1428±.0006
BEN	LWRWIMS	WIMS	-	1.2054	1.1623	1.1427
CEA3	APOLLO2	JEF22CEA	1.3746	1.2131	1.1711	1.1496
CEA4	ECCO52	JEF22CEA	1.3697	1.2107	1.1674	1.1447
CEA5	APOLLO1	ENDFBCEA	1.3630	1.2092	1.1634	1.1391
CEN	DOT35	MOLBR2	1.3731	1.2255	1.1952	1.1835
ECN	MCNP4	JEF22	1.3746±.0005	1.2166±.0007	1.1725±.0008	1.1504±.0009
ENEA	MCNP4	JEF1	1.3608±.0006	1.2131±.0008	1.1676±.0008	1.1435±.0008
HIT	VMONT	JENDL2	1.3582	1.2237	1.1799	1.1549
IKE1	CGM	JEF1	1.3633	1.2137	1.1710	1.1488
IKE2	MCNP4	JEF22	1.3726±.0003	-	-	-
JAE1	SRAC	JENDL31	1.3618	1.2213	1.1782	1.1544
JAE2	MVP	JENDL31	1.3622±.0003	1.2185±.0004	1.1757±.0004	1.1533±.0004
JAE3	SRACPIK	JENDL31	1.3618	1.2241	1.1782	1.1544
JAE4	MOSRA	JENDL31	1.3618	1.2213	1.1782	1.1544
TOS	MCNP42	JENDL31	1.3609±.0010	1.2165±.0011	1.1715±.0009	1.1501±.0012
Average			1.3656	1.2163	1.1733	1.1511

Table 4 : k-infinity, Pincell moderated, c/a-values

Contr.	Code	Data	Fuel Type			
			UO_2	H-MOX	M-MOX	L-MOX
ANL	VIM	ENDFB5	0.9997±.0007	0.9968±.0010	0.9947±.0009	0.9928±.0005
BEN	LWRWIMS	WIMS	-	0.9910	0.9906	0.9927
CEA3	APOLLO2	JEF22CEA	1.0066	0.9973	0.9981	0.9987
CEA4	ECCO52	JEF22CEA	1.0030	0.9954	0.9950	0.9944
CEA5	APOLLO1	ENDFBCEA	0.9981	0.9941	0.9916	0.9896
CEN	DOT35	MOLBR2	1.0055	1.0075	1.0187	1.0281
ECN	MCNP4	JEF22	1.0066±.0004	1.0002±.0006	0.9993±.0007	0.9994±.0008
ENEA	MCNP4	JEF1	0.9965±.0004	0.9973±.0007	0.9952±.0007	0.9934±.0007
HIT	VMONT	JENDL2	0.9946	1.0060	1.0056	1.0033
IKE1	CGM	JEF1	0.9984	0.9978	0.9981	0.9980
IKE2	MCNP4	JEF22	1.0052±.0002	-	-	-
JAE1	SRAC	JENDL31	0.9972	1.0041	1.0042	1.0029
JAE2	MVP	JENDL31	0.9975±.0002	1.0018±.0003	1.0021±.0003	1.0019±.0003
JAE3	SRACPIK	JENDL31	0.9972	1.0064	1.0042	1.0029
JAE4	MOSRA	JENDL31	0.9972	1.0041	1.0042	1.0029
TOS	MCNP42	JENDL31	0.9966±.0007	1.0001±.0009	0.9985±.0008	0.9991±.0010

Table 5 : k-infinity, Pincell voided

Contr.	Code	Data	Fuel Type			
			UO_2	H-MOX	M-MOX	L-MOX
ANL	VIM	ENDFB5	0.6215±.0006	1.2850±.0007	1.0380±.0006	0.7616±.0011
BEN	LWRWIMS	WIMS	-	1.2592	1.0217	0.7574
CEA3	APOLLO2	JEF22CEA	0.6500	1.2879	1.0475	0.7766
CEA4	ECCO52	JEF22CEA	0.6444	1.2850	1.0441	0.7738
CEA5	APOLLO1	ENDFBCEA	0.6234	1.2860	1.0398	0.7637
CEN	DOT35	MOLBR2	0.6211	1.2465	1.0051	0.7379
ECN	MCNP4	JEF22	0.6380±.0006	1.2863±.0008	1.0427±.0004	0.7696±.0005
ENEA	MCNP4	JEF1	-	1.2668±.0011	1.0216±.0013	0.7495±.0018
HIT	VMONT	JENDL2	0.6211	1.2781	1.0324	0.7595
IKE1	CGM	JEF1	0.6212	1.2712	1.0284	0.7573
IKE2	MCNP4	JEF22	-	-	-	-
JAE1	SRAC	JENDL31	0.6264	1.2741	1.0331	0.7650
JAE2	MVP	JENDL31	0.6228±.0003	1.2689±.0004	1.0280±.0003	0.7598±.0005
JAE3	SRACPIK	JENDL31	0.6264	1.2741	1.0331	0.7651
JAE4	MOSRA	JENDL31	0.6264	1.2741	1.0331	0.7650
TOS	MCNP42	JENDL31	0.6236±.0013	1.2783±.0013	1.0369±.0013	0.7657±.0011
Average			0.6282	1.2748	1.0324	0.7618

Table 6 : k-infinity, Pincell voided, c/a-values

Contr.	Code	Data	Fuel Type			
			UO_2	H-MOX	M-MOX	L-MOX
ANL	VIM	ENDFB5	0.9894±.0010	1.0080±.0005	1.0055±.0006	0.9997±.0014
BEN	LWRWIMS	WIMS	-	0.9878	0.9896	0.9942
CEA3	APOLLO2	JEF22CEA	1.0347	1.0103	1.0147	1.0194
CEA4	ECCO52	JEF22CEA	1.0259	1.0081	1.0114	1.0156
CEA5	APOLLO1	ENDFBCEA	0.9924	1.0088	1.0072	1.0024
CEN	DOT35	MOLBR2	0.9887	0.9778	0.9736	0.9686
ECN	MCNP4	JEF22	1.0156±.0010	1.0090±.0006	1.0100±.0004	1.0102±.0007
ENEA	MCNP4	JEF1	-	0.9938±.0009	0.9896±.0013	0.9839±.0024
HIT	VMONT	JENDL2	0.9887	1.0026	1.0000	0.9969
IKE1	CGM	JEF1	0.9888	0.9972	0.9961	0.9940
IKE2	MCNP4	JEF22	-	-	-	-
JAE1	SRAC	JENDL31	0.9972	0.9995	1.0007	1.0042
JAE2	MVP	JENDL31	0.9914±.0005	0.9954±.0003	0.9958±.0003	0.9973±.0007
JAE3	SRACPIK	JENDL31	0.9972	0.9995	1.0007	1.0043
JAE4	MOSRA	JENDL31	0.9972	0.9995	1.0007	1.0042
TOS	MCNP42	JENDL31	0.9928±.0021	1.0027±.0010	1.0044±.0013	1.0050±.0014

Table 7 : k-infinity, Macrocell moderated

Contr.	Code	Data	Fuel Type			
			UO_2	H-MOX	M-MOX	L-MOX
ANL	VIM	ENDFB5	1.3653±.0002	1.3428±.0002	1.3391±.0002	1.3382±.0002
BEN	LWRWIMS	WIMS	-	1.3452	1.3418	1.3415
CEA3	APOLLO2	JEF22CEA	1.3745	1.3508	1.3472	1.3464
CEA4	ECCO52	JEF22CEA	1.3697	1.3464	1.3428	1.3417
CEA5	APOLLO1	ENDFBCEA	1.3630	1.3403	1.3363	1.3353
CEN	DOT35	MOLBR2	1.3773	1.3561	1.3537	1.3538
ECN	MCNP4	JEF22	1.3773±.0010	1.3516±.0007	1.3495±.0007	1.3481±.0008
ENEA	MCNP4	JEF1	1.3600±.0007	1.3228±.0007	1.3201±.0008	1.3198±.0007
HIT	VMONT	JENDL2	1.3577	1.3376	1.3341	1.3324
IKE1	CGM	JEF1	1.3657	1.3438	1.3387	1.3392
IKE2	MCNP4	JEF22	1.3726±.0003	1.3487±.0003	1.3446±.0003	1.3440±.0003
JAE1	SRAC	JENDL31	1.3617	1.3404	1.3367	1.3357
JAE2	MVP	JENDL31	1.3629±.0003	1.3416±.0003	1.3383±.0003	1.3374±.0003
JAE3	SRACPIK	JENDL31	1.3618	1.3411	1.3373	1.3361
JAE4	MOSRA	JENDL31	1.3615	1.3403	1.3367	1.3356
TOS	MCNP42	JENDL31	1.3601±.0010	1.3382±.0009	1.3364±.0009	1.3334±.0009
Average			1.3661	1.3430	1.3396	1.3387

Table 8 : k-infinity, Macrocell moderated, c/a-values

Contr.	Code	Data	Fuel Type			
			UO_2	H-MOX	M-MOX	L-MOX
ANL	VIM	ENDFB5	0.9994±.0001	0.9999±.0001	0.9996±.0001	0.9997±.0001
BEN	LWRWIMS	WIMS	-	1.0016	1.0016	1.0021
CEA3	APOLLO2	JEF22CEA	1.0062	1.0058	1.0057	1.0058
CEA4	ECCO52	JEF22CEA	1.0026	1.0026	1.0024	1.0022
CEA5	APOLLO1	ENDFBCEA	0.9978	0.9980	0.9976	0.9975
CEN	DOT35	MOLBR2	1.0082	1.0098	1.0106	1.0113
ECN	MCNP4	JEF22	1.0082±.0007	1.0064±.0005	1.0074±.0005	1.0071±.0006
ENEA	MCNP4	JEF1	0.9956±.0005	0.9850±.0005	0.9855±.0006	0.9859±.0005
HIT	VMONT	JENDL2	0.9939	0.9960	0.9959	0.9953
IKE1	CGM	JEF1	0.9997	1.0006	0.9993	1.0004
IKE2	MCNP4	JEF22	1.0048±.0002	1.0043±.0002	1.0037±.0002	1.0040±.0002
JAE1	SRAC	JENDL31	0.9968	0.9981	0.9979	0.9978
JAE2	MVP	JENDL31	0.9977±.0002	0.9990±.0002	0.9991±.0002	0.9991±.0002
JAE3	SRACPIK	JENDL31	0.9969	0.9986	0.9983	0.9981
JAE4	MOSRA	JENDL31	0.9967	0.9980	0.9978	0.9977
TOS	MCNP42	JENDL31	0.9956±.0007	0.9964±.0007	0.9976±.0007	0.9960±.0007

Table 9 : k-infinity, Macrocell voided

Contr.	Code	Data	Fuel Type			
			UO_2	H-MOX	M-MOX	L-MOX
ANL	VIM	ENDFB5	1.3508±.0002	1.3481±.0002	1.3434±.0002	1.3398±.0002
BEN	LWRWIMS	WIMS	-	1.3513	1.3469	1.3440
CEA3	APOLLO2	JEF22CEA	1.3603	1.3559	1.3510	1.3475
CEA4	ECCO52	JEF22CEA	1.3557	1.3507	1.3461	1.3430
CEA5	APOLLO1	ENDFBCEA	1.3492	1.3460	1.3410	1.3375
CEN	DOT35	MOLBR2	1.3609	1.3584	1.3537	1.3507
ECN	MCNP4	JEF22	1.3618±.0005	1.3583±.0005	1.3547±.0008	1.3505±.0007
ENEA	MCNP4	JEF1	1.3454	1.3248±.0007	1.3196±.0008	1.3155±.0007
HIT	VMONT	JENDL2	1.3434	1.3419	1.3371	1.3330
IKE1	CGM	JEF1	1.3501	1.3455	1.3408	1.3377
IKE2	MCNP4	JEF22	1.3588±.0003	1.3549±.0003	1.3502±.0003	1.3466±.0003
JAE1	SRAC	JENDL31	1.3460	1.3408	1.3363	1.3332
JAE2	MVP	JENDL31	1.3474±.0003	1.3457±.0003	1.3415±.0004	1.3372±.0003
JAE3	SRACPIK	JENDL31	1.3468	1.3450	1.3401	1.3364
JAE4	MOSRA	JENDL31	1.3459	1.3428	1.3379	1.3344
TOS	MCNP42	JENDL31	1.3450±.0009	1.3422±.0009	1.3390±.0009	1.3356±.0009
Average			1.3512	1.3470	1.3425	1.3389

Table 10 : k-infinity, Macrocell voided c/a-values

Contr.	Code	Data	Fuel Type			
			UO_2	H-MOX	M-MOX	L-MOX
ANL	VIM	ENDFB5	0.9997±.0001	1.0008±.0001	1.0007±.0001	1.0007±.0001
BEN	LWRWIMS	WIMS	-	1.0032	1.0033	1.0038
CEA3	APOLLO2	JEF22CEA	1.0068	1.0066	1.0064	1.0064
CEA4	ECCO52	JEF22CEA	1.0033	1.0027	1.0027	1.0031
CEA5	APOLLO1	ENDFBCEA	0.9985	0.9992	0.9989	0.9989
CEN	DOT35	MOLBR2	1.0072	1.0085	1.0084	1.0088
ECN	MCNP4	JEF22	1.0079±.0004	1.0084±.0004	1.0091±.0006	1.0086±.0005
ENEA	MCNP4	JEF1	0.9957	0.9835±.0005	0.9830±.0006	0.9825±.0005
HIT	VMONT	JENDL2	0.9943	0.9962	0.9960	0.9956
IKE1	CGM	JEF1	0.9992	0.9989	0.9988	0.9991
IKE2	MCNP4	JEF22	1.0057±.0002	1.0059±.0002	1.0058±.0002	1.0058±.0002
JAE1	SRAC	JENDL31	0.9962	0.9954	0.9954	0.9957
JAE2	MVP	JENDL31	0.9972±.0002	0.9990±.0002	0.9993±.0003	0.9987±.0002
JAE3	SRACPIK	JENDL31	0.9968	0.9985	0.9982	0.9981
JAE4	MOSRA	JENDL31	0.9961	0.9968	0.9966	0.9966
TOS	MCNP42	JENDL31	0.9955±.0007	0.9964±.0007	0.9974±.0007	0.9976±.0007

Table 11 : Delta-k-Void, Pincell

Contr.	Code	Data	Fuel Type			
			UO_2	H-MOX	M-MOX	L-MOX
ANL	VIM	ENDFB5	-0.7436±.0016	0.0726±.0019	-0.1291±.0017	-0.3812±.0017
BEN	LWRWIMS	WIMS	-	0.0538	-0.1406	-0.3853
CEA3	APOLLO2	JEF22CEA	-0.7246	0.0748	-0.1236	-0.3730
CEA4	ECCO52	JEF22CEA	-0.7252	0.0743	-0.1233	-0.3709
CEA5	APOLLO1	ENDFBCEA	-0.7396	0.0768	-0.1236	-0.3754
CEN	DOT35	MOLBR2	-0.7520	0.0210	-0.1901	-0.4456
ECN	MCNP4	JEF22	-0.7366±.0011	0.0697±.0015	-0.1298±.0012	-0.3808±.0014
ENEA	MCNP4	JEF1	-	0.0537±.0019	-0.1460±.0021	-0.3940±.0026
HIT	VMONT	JENDL2	-0.7371	0.0544	-0.1475	-0.3954
IKE1	CGM	JEF1	-0.7422	0.0575	-0.1427	-0.3915
IKE2	MCNP4	JEF22	-	-	-	-
JAE1	SRAC	JENDL31	-0.7354	0.0527	-0.1451	-0.3894
JAE2	MVP	JENDL31	-0.7394±.0006	0.0504±.0008	-0.1477±.0007	-0.3935±.0009
JAE3	SRACPIK	JENDL31	-0.7354	0.0500	-0.1451	-0.3893
JAE4	MOSRA	JENDL31	-0.7354	0.0527	-0.1451	-0.3894
TOS	MCNP42	JENDL31	-0.7373±.0023	0.0618±.0024	-0.1346±.0022	-0.3844±.0023
Average			-0.7372	0.0584	-0.1409	-0.3893

Table 12 : Delta-k-Void, Pincell, c/a-values

Contributor	Code	Data	Fuel Type			
			UO_2	H-MOX	M-MOX	L-MOX
ANL	VIM	ENDFB5	1.0087	1.2427	0.9161	0.9793
BEN	LWRWIMS	WIMS	0.0000	0.9214	0.9977	0.9899
CEA3	APOLLO2	JEF22CEA	0.9829	1.2804	0.8771	0.9582
CEA4	ECCO52	JEF22CEA	0.9837	1.2723	0.8749	0.9528
CEA5	APOLLO1	ENDFBCEA	1.0032	1.3146	0.8771	0.9644
CEN	DOT35	MOLBR2	1.0201	0.3603	1.3490	1.1446
ECN	MCNP4	JEF22	0.9992	1.1931	0.9211	0.9782
ENEA	MCNP4	JEF1	0.0000	0.9193	1.0362	1.0122
HIT	VMONT	JENDL2	0.9998	0.9312	1.0467	1.0157
IKE1	CGM	JEF1	1.0067	0.9839	1.0124	1.0058
IKE2	MCNP4	JEF22	0.0000	0.0000	0.0000	0.0000
JAE1	SRAC	JENDL31	0.9975	0.9024	1.0294	1.0002
JAE2	MVP	JENDL31	1.0030	0.8627	1.0481	1.0109
JAE3	SRACPIK	JENDL31	0.9975	0.8559	1.0296	1.0001
JAE4	MOSRA	JENDL31	0.9975	0.9024	1.0294	1.0002
TOS	MCNP42	JENDL31	1.0001	1.0572	0.9554	0.9876

Table 13 : Delta-k-Void, Macrocell

Contr.	Code	Data	Fuel Type			
			UO_2	H-MOX	M-MOX	L-MOX
ANL	VIM	ENDFB5	-0.0145±.0004	0.0053±.0004	0.0043±.0004	0.0016±.0004
BEN	LWRWIMS	WIMS	-	0.0062	0.0051	0.0026
CEA3	APOLLO2	JEF22CEA	-0.0142	0.0051	0.0038	0.0011
CEA4	ECCO52	JEF22CEA	-0.0140	0.0042	0.0034	0.0014
CEA5	APOLLO1	ENDFBCEA	-0.0138	0.0057	0.0047	0.0022
CEN	DOT35	MOLBR2	-0.0164	0.0023	0.0000	-0.0031
ECN	MCNP4	JEF22	-0.0155±.0015	0.0067±.0012	0.0052±.0015	0.0024±.0015
ENEA	MCNP4	JEF1	-0.0146±.0007	0.0020±.0014	-0.0005±.0016	-0.0043±.0014
HIT	VMONT	JENDL2	-0.0143	0.0043	0.0030	0.0006
IKE1	CGM	JEF1	-0.0156	0.0017	0.0021	-0.0015
IKE2	MCNP4	JEF22	-0.0138±.0006	0.0062±.0006	0.0057±.0006	0.0026±.0006
JAE1	SRAC	JENDL31	-0.0157	0.0004	-0.0004	-0.0025
JAE2	MVP	JENDL31	-0.0155±.0006	0.0041±.0006	0.0032±.0007	-0.0002±.0006
JAE3	SRACPIK	JENDL31	-0.0150	0.0039	0.0028	0.0003
JAE4	MOSRA	JENDL31	-0.0156	0.0025	0.0013	-0.0012
TOS	MCNP42	JENDL31	-0.0151±.0019	0.0040±.0018	0.0026±.0018	0.0023±.0018
Average			-0.0149	0.0040	0.0029	0.0003

Table 14 : Delta-k-Void, Macrocell, c/a-values

Contributor	Code	Data	Fuel Type			
			UO_2	H-MOX	M-MOX	L-MOX
ANL	VIM	ENDFB5	0.9730	1.3121	1.4898	6.0924
BEN	LWRWIMS	WIMS	0.0000	1.5275	1.7739	9.7097
CEA3	APOLLO2	JEF22CEA	0.9528	1.2626	1.3165	4.1883
CEA4	ECCO52	JEF22CEA	0.9394	1.0497	1.1676	5.1782
CEA5	APOLLO1	ENDFBCEA	0.9260	1.4111	1.6284	8.3770
CEN	DOT35	MOLBR2	1.0998	0.5744	0.0035	-11.917
ECN	MCNP4	JEF22	1.0401	1.6587	1.8016	9.1386
ENEA	MCNP4	JEF1	0.9806	0.4964	-0.1781	-16.530
HIT	VMONT	JENDL2	0.9595	1.0645	1.0394	2.2845
IKE1	CGM	JEF1	1.0474	0.4208	0.7379	-5.7493
IKE2	MCNP4	JEF22	0.9266	1.5376	1.9606	9.9952
JAE1	SRAC	JENDL31	1.0515	0.0990	-0.1490	-9.5195
JAE2	MVP	JENDL31	1.0401	1.0150	1.1087	-0.7612
JAE3	SRACPIK	JENDL31	1.0038	0.9581	0.9563	1.0281
JAE4	MOSRA	JENDL31	1.0488	0.6140	0.4366	-4.4933
TOS	MCNP42	JENDL31	1.0107	0.9985	0.9063	8.6852

Table 15 : Fission Density AM-Traverse, UOX, Moderated

Contr.	Pin-Position				
	1.0	2.0	3.0	4.0	5.0
ANL	1.000±.004	1.001±.005	1.003±.004	1.003±.005	1.003±.005
BEN	1.000	1.000	1.000	1.000	1.000
CEA3	1.000	1.000	1.000	1.000	1.000
CEA4	1.000	1.000	1.000	1.000	1.000
CEA5	1.000	1.000	1.000	1.000	1.000
CEN	1.000	1.000	1.000	1.000	1.000
ECN	1.000	1.020±.012	1.038±.012	1.036±.012	1.036±.012
HIT	1.000	1.005	1.003	1.001	1.003
IKE1	1.000	1.000	1.000	1.000	1.000
IKE2	1.000±.005	0.999±.004	1.002±.004	0.994±.003	0.997±.003
JAE1	1.000	0.999	0.999	0.999	0.999
JAE2	1.000±.009	0.996±.006	1.005±.006	1.015±.006	1.008±.005
JAE3	1.000	1.000	1.000	0.999	0.999
JAE4	1.000	1.000	1.000	1.000	1.000
TOS	1.000±.020	0.954±.013	0.981±.013	0.973±.012	0.990±.012
Average	1.000	0.998	1.002	1.001	1.002

Contr.	Pin-Position				
	6.0	7.0	8.0	9.0	10.0
ANL	1.004±.004	1.004±.005	1.000±.005	1.000±.005	1.002±.005
BEN	1.000	1.000	1.000	1.000	1.000
CEA3	1.000	1.000	1.000	1.000	1.000
CEA4	1.000	1.000	1.000	1.000	1.000
CEA5	1.000	1.000	1.000	1.000	1.000
CEN	1.000	1.000	1.000	1.000	1.000
ECN	1.010±.012	1.000±.013	1.019±.012	1.001±.013	1.007±.013
HIT	1.002	0.999	0.998	0.999	0.999
IKE1	1.000	1.000	1.000	1.000	1.000
IKE2	1.001±.004	1.002±.004	1.005±.004	1.008±.004	1.001±.004
JAE1	0.999	0.999	0.999	0.999	0.999
JAE2	1.015±.005	1.010±.007	1.011±.005	1.011±.005	1.010±.006
JAE3	0.999	1.000	1.000	1.000	1.000
JAE4	1.000	1.000	1.000	1.000	1.000
TOS	0.977±.012	1.003±.013	0.992±.012	0.974±.012	0.984±.012
Average	1.001	1.001	1.002	0.999	1.000

Contr.	Pin-Position				
	11.0	12.0	13.0	14.0	15.0
ANL	1.002±.005	1.006±.005	1.005±.005	1.005±.005	1.003±.005
BEN	1.000	1.000	1.000	1.000	1.000
CEA3	1.000	1.000	1.000	1.000	1.000
CEA4	1.000	1.000	1.000	1.000	1.000
CEA5	1.000	1.000	1.000	1.000	1.000
CEN	1.000	1.000	1.000	1.000	1.000
ECN	1.003±.012	1.025±.012	1.037±.013	1.024±.013	1.015±.014
HIT	0.996	0.997	0.992	0.999	1.000
IKE1	1.000	1.000	1.000	1.000	1.000
IKE2	1.008±.004	1.007±.004	1.011±.004	1.009±.004	1.003±.004
JAE1	0.999	0.999	0.999	0.999	0.999
JAE2	1.008±.007	1.019±.007	1.010±.006	1.013±.006	1.024±.006
JAE3	1.000	1.000	1.000	1.000	1.000
JAE4	1.000	1.000	1.000	1.000	1.000
TOS	0.997±.013	0.992±.012	1.017±.013	0.980±.012	0.971±.013
Average	1.001	1.003	1.005	1.002	1.001

Table 16 : Fission Density AM-Traverse, UOX, Voided

Contr.	Pin-Position				
	1.0	2.0	3.0	4.0	5.0
ANL	1.000±.002	1.030±.004	1.089±.004	1.187±.004	1.388±.007
BEN	-	-	-	-	-
CEA3	1.000	1.023	1.083	1.188	1.372
CEA4	1.000	1.029	1.085	1.194	1.419
CEA5	1.000	1.029	1.087	1.201	1.442
CEN	1.000	1.027	1.086	1.194	1.401
ECN	1.000	1.009±.012	1.080±.014	1.189±.014	1.401±.018
HIT	1.000	1.020	1.076	1.179	1.375
IKE1	1.000	1.029	1.091	1.199	1.376
IKE2	1.000±.005	1.024±.004	1.088±.004	1.196±.004	1.386±.005
JAE1	1.000	1.027	1.093	1.208	1.397
JAE2	1.000±.010	1.027±.007	1.071±.007	1.183±.008	1.380±.010
JAE3	1.000	1.028	1.092	1.210	1.444
JAE4	1.000	1.031	1.100	1.222	1.428
TOS	1.000±.019	1.045±.015	1.109±.015	1.213±.016	1.429±.018
Average	1.000	1.027	1.088	1.197	1.403

Contr.	Pin-Position				
	6.0	7.0	8.0	9.0	10.0
ANL	1.995±.015	2.588±.021	2.889±.027	3.050±.028	3.151±.033
BEN	-	-	-	-	-
CEA3	1.881	2.346	2.616	2.774	2.866
CEA4	1.977	2.629	2.918	3.091	3.181
CEA5	2.038	2.723	3.021	3.192	3.278
CEN	2.096	2.712	3.014	3.179	3.270
ECN	1.976±.026	2.606±.034	2.903±.035	3.076±.037	3.093±.037
HIT	1.965	2.541	2.846	2.994	3.094
IKE1	1.992	2.572	2.871	3.035	3.126
IKE2	2.000±.008	2.597±.009	2.894±.010	3.043±.010	3.129±.011
JAE1	1.901	2.474	2.777	2.944	3.037
JAE2	2.005±.014	2.592±.018	2.903±.017	3.059±.018	3.147±.020
JAE3	2.049	2.641	2.938	3.087	3.175
JAE4	1.958	2.572	2.892	3.064	3.157
TOS	2.056±.028	2.667±.034	2.999±.037	3.221±.039	3.254±.039
Average	1.992	2.590	2.891	3.058	3.140

Contr.	Pin-Position				
	11.0	12.0	13.0	14.0	15.0
ANL	3.209±.030	3.235±.038	3.252±.034	3.251±.034	3.256±.036
BEN	-	-	-	-	-
CEA3	2.918	2.948	2.964	2.972	2.975
CEA4	3.232	3.257	3.271	3.277	3.279
CEA5	3.326	3.350	3.362	3.368	3.370
CEN	3.320	3.348	3.363	3.371	3.374
ECN	3.128±.038	3.141±.038	3.165±.038	3.171±.038	3.228±.042
HIT	3.143	3.155	3.164	3.162	3.171
IKE1	3.176	3.202	3.215	3.221	3.223
IKE2	3.194±.011	3.227±.011	3.241±.011	3.249±.011	3.234±.012
JAE1	3.088	3.115	3.129	3.135	3.138
JAE2	3.182±.020	3.200±.017	3.228±.018	3.229±.021	3.257±.022
JAE3	3.226	3.254	3.270	3.278	3.282
JAE4	3.207	3.232	3.245	3.250	3.252
TOS	3.251±.039	3.281±.039	3.267±.038	3.282±.039	3.258±.043
Average	3.186	3.210	3.224	3.230	3.236

Table 17 : Fission Density AD-Traverse, UOX, Moderated

Contr.	Pin-Position				
	1.0	2.0	3.0	4.0	5.0
ANL	1.000±.004	1.005±.006	1.002±.005	1.005±.005	1.008±.005
BEN	1.000	1.000	1.000	1.000	1.000
CEA3	1.000	1.000	1.000	1.000	1.000
CEA4	1.000	1.000	1.000	1.000	1.000
CEA5	1.000	1.000	1.000	1.000	1.000
CEN	1.000	1.000	1.000	1.000	1.000
ECN	1.000	1.046±.013	1.021±.011	1.010±.012	1.024±.011
HIT	1.000	0.999	0.997	0.998	0.997
IKE1	1.000	1.000	1.000	1.000	1.000
IKE2	1.000±.005	1.004±.005	1.003±.005	1.010±.005	1.003±.005
JAE1	1.000	0.999	0.999	0.999	0.999
JAE2	1.000±.009	1.001±.008	1.024±.008	1.011±.008	1.005±.008
JAE3	1.000	1.000	0.999	0.999	0.999
JAE4	1.000	1.000	1.000	1.000	1.000
TOS	1.000±.020	0.991±.017	0.994±.016	1.020±.016	1.004±.016
Average	1.000	1.003	1.003	1.004	1.003

Contr.	Pin-Position				
	6.0	7.0	8.0	9.0	10.0
ANL	1.002±.006	1.003±.005	0.998±.005	1.001±.005	1.002±.005
BEN	1.000	1.000	1.000	1.000	1.000
CEA3	1.000	1.000	1.000	1.000	1.000
CEA4	1.000	1.000	1.000	1.000	1.000
CEA5	1.000	1.000	1.000	1.000	1.000
CEN	1.000	1.000	1.000	1.000	1.000
ECN	1.035±.011	1.017±.011	1.023±.011	1.023±.011	1.045±.011
HIT	1.000	0.995	0.992	1.000	0.993
IKE1	1.000	1.000	1.000	1.000	1.000
IKE2	1.005±.005	1.002±.005	1.001±.005	0.998±.004	1.002±.005
JAE1	0.999	0.999	0.999	0.999	0.999
JAE2	1.016±.008	1.013±.008	1.019±.008	1.021±.009	1.011±.008
JAE3	0.999	0.999	1.000	1.000	1.000
JAE4	1.000	1.000	1.000	1.000	1.000
TOS	0.991±.016	0.994±.016	0.995±.016	1.024±.016	0.972±.016
Average	1.003	1.002	1.002	1.004	1.002

Contr.	Pin-Position				
	11.0	12.0	13.0	14.0	15.0
ANL	1.004±.005	1.002±.005	1.002±.006	1.007±.005	1.000±.006
BEN	1.000	1.000	1.000	1.000	1.000
CEA3	1.000	1.000	1.000	1.000	1.000
CEA4	1.000	1.000	1.000	1.000	1.000
CEA5	1.000	1.000	1.000	1.000	1.000
CEN	1.000	1.000	1.000	1.000	1.000
ECN	1.023±.011	1.025±.011	1.026±.012	1.033±.012	1.013±.014
HIT	1.001	1.000	0.998	1.000	0.998
IKE1	1.000	1.000	1.000	1.000	1.000
IKE2	1.002±.005	1.006±.005	1.004±.005	1.010±.005	1.001±.006
JAE1	0.999	0.999	0.999	0.999	0.999
JAE2	1.021±.008	1.021±.008	1.030±.008	1.017±.008	1.005±.010
JAE3	1.000	1.000	1.000	1.000	1.000
JAE4	1.000	1.000	1.000	1.000	1.000
TOS	0.996±.016	1.007±.016	0.985±.016	0.986±.016	1.055±.020
Average	1.003	1.004	1.003	1.003	1.005

Table 18 : Fission Density AD-Traverse, UOX, Voided

Contr.	Pin-Position				
	1.0	2.0	3.0	4.0	5.0
ANL	1.000±.002	1.055±.005	1.166±.005	1.384±.006	1.786±.012
BEN	-	-	-	-	-
CEA3	1.000	1.055	1.174	1.382	1.752
CEA4	1.000	1.052	1.170	1.392	1.857
CEA5	1.000	1.053	1.175	1.406	1.906
CEN	1.000	1.054	1.173	1.390	1.813
ECN	1.000	1.049±.013	1.178±.013	1.365±.015	1.757±.019
HIT	1.000	1.049	1.159	1.368	1.769
IKE1	1.000	1.057	1.181	1.392	1.744
IKE2	1.000±.005	1.046±.005	1.177±.005	1.387±.006	1.788±.008
JAE1	1.000	1.057	1.188	1.413	1.789
JAE2	1.000±.010	1.054±.009	1.166±.009	1.361±.010	1.777±.014
JAE3	1.000	1.057	1.183	1.416	1.884
JAE4	1.000	1.062	1.199	1.438	1.843
TOS	1.000±.019	1.054±.017	1.196±.019	1.430±.023	1.833±.030
Average	1.000	1.054	1.177	1.395	1.807

Contr.	Pin-Position				
	6.0	7.0	8.0	9.0	10.0
ANL	2.659±.026	3.040±.031	3.186±.039	3.250±.042	3.246±.036
BEN	-	-	-	-	-
CEA3	2.456	2.775	2.903	2.957	2.980
CEA4	2.632	3.050	3.197	3.252	3.272
CEA5	2.721	3.147	3.294	3.346	3.363
CEN	2.757	3.160	3.295	3.348	3.371
ECN	2.592±.031	2.994±.033	3.189±.035	3.175±.035	3.164±.035
HIT	2.607	2.985	3.126	3.168	3.174
IKE1	2.652	3.026	3.154	3.201	3.217
IKE2	2.653±.012	3.041±.013	3.174±.014	3.227±.014	3.231±.014
JAE1	2.543	2.930	3.063	3.113	3.131
JAE2	2.636±.024	3.008±.018	3.162±.023	3.174±.025	3.238±.026
JAE3	2.691	3.080	3.205	3.254	3.275
JAE4	2.645	3.047	3.181	3.229	3.246
TOS	2.723±.045	3.122±.049	3.243±.049	3.375±.052	3.389±.052
Average	2.641	3.029	3.170	3.219	3.236

Contr.	Pin-Position				
	11.0	12.0	13.0	14.0	15.0
ANL	3.241±.038	3.259±.042	3.257±.039	3.257±.035	3.260±.041
BEN	-	-	-	-	-
CEA3	2.988	2.992	2.993	2.993	2.993
CEA4	3.278	3.280	3.280	3.279	3.279
CEA5	3.368	3.369	3.367	3.367	3.367
CEN	3.381	3.386	3.387	3.390	3.390
ECN	3.209±.035	3.221±.035	3.145±.035	3.225±.035	3.216±.042
HIT	3.188	3.175	3.210	3.182	3.207
IKE1	3.222	3.223	3.222	3.222	3.222
IKE2	3.243±.014	3.223±.014	3.242±.014	3.233±.015	3.203±.017
JAE1	3.138	3.139	3.139	3.139	3.139
JAE2	3.283±.022	3.253±.026	3.276±.024	3.229±.025	3.261±.033
JAE3	3.283	3.286	3.287	3.287	3.287
JAE4	3.250	3.251	3.251	3.250	3.250
TOS	3.372±.052	3.338±.051	3.381±.052	3.323±.052	3.382±.064
Average	3.246	3.242	3.246	3.241	3.247

Table 19 : Fission Density AM-Traverse, H-MOX, Moderated

Contr.	Pin-Position				
	1.0	2.0	3.0	4.0	5.0
ANL	1.000±.003	0.996±.004	1.025±.004	1.113±.005	1.544±.008
BEN	1.000	1.008	1.034	1.123	1.549
CEA3	1.000	1.014	1.050	1.133	1.327
CEA4	1.000	1.002	1.036	1.081	1.623
CEA5	1.000	1.002	1.033	1.076	1.606
CEN	1.000	1.008	1.036	1.131	1.550
ECN	1.000	1.003±.016	1.044±.018	1.112±.018	1.560±.023
HIT	1.000	1.005	1.027	1.114	1.528
IKE1	1.000	1.009	1.039	1.138	1.540
IKE2	1.000±.007	1.000±.005	1.033±.005	1.122±.005	1.549±.006
JAE1	1.000	1.006	1.034	1.133	1.565
JAE2	1.000±.014	1.025±.008	1.032±.007	1.123±.008	1.576±.011
JAE3	1.000	1.009	1.036	1.135	1.529
JAE4	1.000	1.008	1.036	1.134	1.557
TOS	1.000±.026	1.039±.018	1.036±.018	1.115±.018	1.575±.023
Average	1.000	1.009	1.035	1.119	1.545
	6.0	7.0	8.0	9.0	10.0
ANL	0.840±.002	1.031±.004	1.137±.004	1.194±.006	1.231±.005
BEN	0.833	1.047	1.160	1.219	1.253
CEA3	1.016	1.143	1.221	1.269	1.298
CEA4	0.813	1.030	1.143	1.209	1.245
CEA5	0.802	1.023	1.136	1.198	1.231
CEN	0.827	1.029	1.136	1.196	1.230
ECN	0.846±.012	1.030±.014	1.133±.015	1.240±.016	1.238±.016
HIT	0.823	1.005	1.106	1.166	1.201
IKE1	0.813	1.010	1.121	1.183	1.219
IKE2	0.816±.003	0.995±.004	1.104±.004	1.158±.004	1.189±.004
JAE1	0.812	1.002	1.113	1.177	1.213
JAE2	0.848±.007	1.033±.007	1.154±.007	1.201±.006	1.233±.007
JAE3	0.806	1.004	1.110	1.164	1.197
JAE4	0.801	1.004	1.120	1.185	1.221
TOS	0.849±.012	1.046±.014	1.166±.015	1.210±.015	1.260±.016
Average	0.836	1.029	1.137	1.198	1.231
	11.0	12.0	13.0	14.0	15.0
ANL	1.251±.006	1.265±.005	1.277±.005	1.277±.006	1.276±.006
BEN	1.272	1.284	1.292	1.296	1.297
CEA3	1.315	1.326	1.333	1.337	1.338
CEA4	1.265	1.277	1.283	1.287	1.289
CEA5	1.250	1.262	1.268	1.271	1.273
CEN	1.249	1.260	1.266	1.270	1.271
ECN	1.268±.016	1.270±.015	1.261±.016	1.276±.017	1.295±.018
HIT	1.215	1.227	1.234	1.233	1.231
IKE1	1.239	1.250	1.257	1.260	1.262
IKE2	1.212±.004	1.228±.004	1.235±.004	1.232±.004	1.237±.005
JAE1	1.234	1.246	1.253	1.257	1.258
JAE2	1.256±.008	1.264±.008	1.271±.007	1.265±.007	1.276±.009
JAE3	1.217	1.228	1.235	1.239	1.241
JAE4	1.241	1.253	1.259	1.262	1.264
TOS	1.269±.016	1.250±.016	1.262±.016	1.290±.017	1.268±.018
Average	1.250	1.259	1.266	1.270	1.272

Table 20 : Fission Density AM-Traverse, H-MOX, Voided

Contr.	Pin-Position				
	1.0	2.0	3.0	4.0	5.0
ANL	1.000±.004	1.034±.005	1.143±.006	1.424±.009	2.404±.029
BEN	1.000	1.038	1.127	1.364	2.220
CEA3	1.000	1.060	1.205	1.502	2.134
CEA4	1.000	1.025	1.100	1.364	2.661
CEA5	1.000	1.019	1.090	1.322	2.667
CEN	1.000	1.028	1.111	1.369	2.777
ECN	1.000	1.059±.014	1.112±.016	1.380±.021	2.377±.038
HIT	1.000	1.031	1.123	1.398	2.351
IKE1	1.000	1.029	1.116	1.391	2.646
IKE2	1.000±.005	1.045±.004	1.152±.005	1.450±.006	2.466±.011
JAE1	1.000	1.026	1.116	1.403	2.699
JAE2	1.000±.009	1.020±.007	1.125±.008	1.419±.011	2.431±.019
JAE3	1.000	1.028	1.109	1.356	2.668
JAE4	1.000	1.024	1.095	1.332	2.698
TOS	1.000±.018	1.022±.014	1.126±.016	1.418±.022	2.385±.037
Average	1.000	1.033	1.123	1.393	2.506

Contr.	Pin-Position				
	6.0	7.0	8.0	9.0	10.0
ANL	1.606±.012	2.358±.023	2.806±.032	3.060±.038	3.230±.042
BEN	1.639	2.405	2.842	3.093	3.244
CEA3	2.171	2.861	3.305	3.586	3.762
CEA4	1.556	2.385	2.849	3.145	3.317
CEA5	1.615	2.459	2.924	3.205	3.363
CEN	1.806	2.639	3.116	3.402	3.571
ECN	1.608±.023	2.410±.034	2.877±.037	3.071±.037	3.286±.039
HIT	1.583	2.295	2.724	2.965	3.124
IKE1	1.661	2.408	2.868	3.151	3.321
IKE2	1.599±.007	2.331±.009	2.774±.010	3.053±.011	3.222±.011
JAE1	1.481	2.234	2.714	3.013	3.193
JAE2	1.605±.011	2.375±.015	2.773±.017	3.053±.019	3.200±.019
JAE3	1.679	2.442	2.882	3.124	3.277
JAE4	1.568	2.342	2.825	3.118	3.290
TOS	1.602±.024	2.323±.032	2.758±.036	2.915±.036	3.202±.039
Average	1.652	2.418	2.869	3.130	3.307

Contr.	Pin-Position				
	11.0	12.0	13.0	14.0	15.0
ANL	3.332±.047	3.393±.049	3.420±.050	3.432±.048	3.438±.053
BEN	3.330	3.381	3.412	3.430	3.436
CEA3	3.868	3.932	3.969	3.989	3.998
CEA4	3.418	3.474	3.503	3.518	3.524
CEA5	3.453	3.505	3.532	3.546	3.551
CEN	3.670	3.727	3.758	3.775	3.782
ECN	3.373±.040	3.376±.041	3.347±.040	3.383±.041	3.348±.044
HIT	3.203	3.253	3.275	3.301	3.301
IKE1	3.420	3.476	3.507	3.522	3.528
IKE2	3.306±.011	3.362±.011	3.399±.011	3.404±.012	3.411±.013
JAE1	3.299	3.359	3.391	3.408	3.415
JAE2	3.312±.019	3.391±.021	3.369±.021	3.415±.022	3.427±.021
JAE3	3.371	3.426	3.457	3.474	3.482
JAE4	3.388	3.442	3.470	3.484	3.489
TOS	3.286±.039	3.317±.039	3.299±.040	3.424±.042	3.366±.045
Average	3.402	3.454	3.474	3.500	3.500

Table 21 : Fission Density AD-Traverse, H-MOX, Moderated

Contr.	Pin-Position				
	1.0	2.0	3.0	4.0	5.0
ANL	1.000±.003	1.013±.005	1.055±.006	1.226±.006	2.008±.015
BEN	1.000	1.016	1.069	1.243	2.026
CEA3	1.000	1.027	1.095	1.245	1.570
CEA4	1.000	1.006	1.069	1.136	2.156
CEA5	1.000	1.008	1.066	1.129	2.125
CEN	1.000	1.016	1.071	1.252	2.023
ECN	1.000	1.014±.016	1.082±.017	1.233±.018	2.013±.026
HIT	1.000	1.016	1.061	1.223	1.981
IKE1	1.000	1.018	1.076	1.267	2.016
IKE2	1.000±.007	1.006±.007	1.059±.007	1.229±.007	2.029±.011
JAE1	1.000	1.014	1.068	1.259	2.056
JAE2	1.000±.014	1.042±.011	1.077±.010	1.249±.012	2.039±.013
JAE3	1.000	1.018	1.071	1.260	1.984
JAE4	1.000	1.016	1.069	1.260	2.053
TOS	1.000±.026	1.036±.024	1.071±.024	1.280±.028	2.006±.037
Average	1.000	1.018	1.071	1.233	2.006

Contr.	Pin-Position				
	6.0	7.0	8.0	9.0	10.0
ANL	1.073±.004	1.199±.006	1.248±.006	1.266±.007	1.285±.006
BEN	1.073	1.219	1.273	1.293	1.305
CEA3	1.189	1.272	1.311	1.330	1.341
CEA4	1.058	1.202	1.257	1.280	1.291
CEA5	1.047	1.190	1.244	1.266	1.276
CEN	1.062	1.195	1.243	1.264	1.274
ECN	1.079±.013	1.186±.014	1.246±.015	1.259±.014	1.269±.014
HIT	1.043	1.170	1.213	1.226	1.237
IKE1	1.047	1.182	1.233	1.254	1.264
IKE2	1.033±.005	1.151±.005	1.216±.006	1.228±.006	1.253±.006
JAE1	1.040	1.175	1.227	1.250	1.260
JAE2	1.072±.009	1.225±.009	1.237±.010	1.266±.011	1.313±.010
JAE3	1.035	1.167	1.212	1.232	1.242
JAE4	1.038	1.181	1.234	1.256	1.266
TOS	1.089±.019	1.204±.020	1.221±.020	1.304±.021	1.294±.021
Average	1.065	1.194	1.241	1.265	1.278

Contr.	Pin-Position				
	11.0	12.0	13.0	14.0	15.0
ANL	1.290±.006	1.287±.006	1.294±.007	1.297±.006	1.294±.006
BEN	1.310	1.314	1.318	1.320	1.321
CEA3	1.347	1.351	1.353	1.354	1.355
CEA4	1.298	1.301	1.304	1.305	1.306
CEA5	1.282	1.285	1.287	1.289	1.289
CEN	1.280	1.283	1.285	1.286	1.287
ECN	1.274±.014	1.303±.014	1.268±.014	1.276±.015	1.321±.018
HIT	1.235	1.247	1.251	1.253	1.247
IKE1	1.269	1.272	1.274	1.275	1.275
IKE2	1.252±.006	1.255±.006	1.264±.006	1.254±.006	1.269±.007
JAE1	1.266	1.269	1.271	1.273	1.273
JAE2	1.302±.010	1.281±.011	1.298±.009	1.300±.011	1.308±.012
JAE3	1.248	1.251	1.253	1.254	1.255
JAE4	1.271	1.274	1.276	1.277	1.278
TOS	1.323±.022	1.275±.021	1.334±.022	1.332±.022	1.306±.026
Average	1.283	1.283	1.289	1.290	1.292

Table 22 : Fission Density AD-Traverse, H-MOX, Voided

Contr.	Pin-Position				
	1.0	2.0	3.0	4.0	5.0
ANL	1.000±.004	1.072±.005	1.274±.009	1.896±.017	3.838±.064
BEN	1.000	1.069	1.262	1.755	3.538
CEA3	1.000	1.118	1.396	1.958	3.171
CEA4	1.000	1.062	1.213	1.679	4.713
CEA5	1.000	1.053	1.183	1.589	4.739
CEN	1.000	1.056	1.222	1.751	4.792
ECN	1.000	1.064±.013	1.241±.017	1.898±.028	3.830±.050
HIT	1.000	1.070	1.262	1.845	3.714
IKE1	1.000	1.058	1.234	1.806	4.574
IKE2	1.000±.005	1.088±.006	1.301±.007	1.924±.012	3.915±.021
JAE1	1.000	1.056	1.237	1.840	4.779
JAE2	1.000±.009	1.069±.009	1.258±.011	1.859±.019	3.793±.034
JAE3	1.000	1.056	1.217	1.718	4.562
JAE4	1.000	1.048	1.190	1.685	4.773
TOS	1.000±.018	1.064±.019	1.266±.026	1.811±.038	3.845±.074
Average	1.000	1.067	1.250	1.801	4.172

Contr.	Pin-Position				
	6.0	7.0	8.0	9.0	10.0
ANL	2.501±.026	3.064±.036	3.306±.051	3.379±.052	3.432±.050
BEN	2.483	3.076	3.300	3.386	3.421
CEA3	3.065	3.581	3.820	3.933	3.987
CEA4	2.438	3.095	3.360	3.469	3.513
CEA5	2.516	3.153	3.405	3.507	3.547
CEN	2.752	3.388	3.628	3.730	3.775
ECN	2.512±.030	3.124±.034	3.261±.036	3.351±.037	3.453±.038
HIT	2.410	2.949	3.164	3.251	3.276
IKE1	2.550	3.145	3.382	3.480	3.519
IKE2	2.486±.012	3.067±.014	3.265±.014	3.363±.015	3.409±.015
JAE1	2.356	2.992	3.250	3.359	3.404
JAE2	2.475±.021	3.057±.024	3.278±.026	3.322±.028	3.400±.027
JAE3	2.537	3.126	3.337	3.429	3.471
JAE4	2.460	3.093	3.340	3.440	3.478
TOS	2.479±.042	2.982±.048	3.159±.050	3.341±.051	3.368±.053
Average	2.535	3.126	3.350	3.449	3.497

Contr.	Pin-Position				
	11.0	12.0	13.0	14.0	15.0
ANL	3.446±.054	3.444±.065	3.445±.060	3.443±.050	3.469±.049
BEN	3.438	3.441	3.445	3.448	3.448
CEA3	4.013	4.025	4.031	4.035	4.036
CEA4	3.532	3.539	3.541	3.542	3.542
CEA5	3.563	3.569	3.570	3.570	3.571
CEN	3.796	3.806	3.811	3.814	3.815
ECN	3.451±.038	3.541±.039	3.515±.039	3.563±.039	3.476±.045
HIT	3.308	3.299	3.313	3.308	3.305
IKE1	3.534	3.539	3.541	3.541	3.542
IKE2	3.440±.015	3.449±.015	3.434±.015	3.455±.015	3.445±.018
JAE1	3.422	3.429	3.431	3.433	3.433
JAE2	3.457±.023	3.393±.024	3.436±.027	3.488±.027	3.434±.033
JAE3	3.489	3.498	3.501	3.502	3.502
JAE4	3.493	3.498	3.499	3.500	3.500
TOS	3.409±.052	3.457±.053	3.459±.053	3.308±.052	3.414±.063
Average	3.519	3.528	3.531	3.530	3.529

Table 23 : Fission Density AM-Traverse, M-MOX, Moderated

Contr.	Pin-Position				
	1.0	2.0	3.0	4.0	5.0
ANL	1.000±.004	1.007±.005	1.037±.005	1.124±.006	1.485±.010
BEN	1.000	1.008	1.037	1.127	1.492
CEA3	1.000	1.014	1.055	1.166	1.483
CEA4	1.000	1.004	1.037	1.098	1.553
CEA5	1.000	1.004	1.035	1.093	1.541
CEN	1.000	1.009	1.039	1.131	1.487
ECN	1.000	1.047±.019	1.052±.018	1.111±.018	1.487±.022
HIT	1.000	1.012	1.038	1.126	1.482
IKE1	1.000	1.010	1.040	1.138	1.485
IKE2	1.000±.007	1.012±.005	1.040±.005	1.130±.005	1.498±.006
JAE1	1.000	1.007	1.036	1.135	1.508
JAE2	1.000±.011	1.002±.008	1.025±.008	1.110±.009	1.477±.009
JAE3	1.000	1.010	1.038	1.133	1.475
JAE4	1.000	1.009	1.037	1.134	1.500
TOS	1.000±.026	0.969±.017	1.007±.018	1.081±.018	1.462±.022
Average	1.000	1.008	1.037	1.122	1.494

Contr.	Pin-Position				
	6.0	7.0	8.0	9.0	10.0
ANL	0.934±.005	1.127±.006	1.233±.007	1.289±.008	1.321±.008
BEN	0.930	1.139	1.247	1.303	1.337
CEA3	0.905	1.077	1.181	1.244	1.281
CEA4	0.909	1.122	1.231	1.293	1.327
CEA5	0.902	1.119	1.228	1.287	1.319
CEN	0.909	1.102	1.204	1.261	1.292
ECN	0.940±.013	1.106±.014	1.203±.016	1.288±.017	1.307±.017
HIT	0.920	1.100	1.202	1.259	1.291
IKE1	0.908	1.100	1.207	1.267	1.300
IKE2	0.912±.004	1.092±.004	1.198±.004	1.253±.005	1.281±.004
JAE1	0.903	1.089	1.196	1.257	1.292
JAE2	0.917±.007	1.090±.007	1.191±.006	1.262±.008	1.290±.008
JAE3	0.900	1.093	1.194	1.246	1.278
JAE4	0.895	1.092	1.204	1.265	1.299
TOS	0.915±.013	1.056±.014	1.197±.016	1.257±.016	1.261±.016
Average	0.913	1.100	1.208	1.269	1.298

Contr.	Pin-Position				
	11.0	12.0	13.0	14.0	15.0
ANL	1.338±.008	1.355±.008	1.358±.008	1.363±.009	1.367±.008
BEN	1.354	1.367	1.374	1.378	1.379
CEA3	1.303	1.316	1.323	1.328	1.330
CEA4	1.346	1.358	1.364	1.368	1.369
CEA5	1.338	1.349	1.355	1.359	1.360
CEN	1.310	1.321	1.327	1.330	1.332
ECN	1.340±.017	1.362±.016	1.346±.016	1.381±.018	1.347±.019
HIT	1.304	1.320	1.325	1.319	1.322
IKE1	1.319	1.330	1.337	1.340	1.342
IKE2	1.307±.005	1.311±.005	1.313±.005	1.325±.005	1.323±.005
JAE1	1.311	1.323	1.329	1.333	1.334
JAE2	1.301±.008	1.324±.008	1.328±.008	1.330±.008	1.326±.010
JAE3	1.296	1.308	1.314	1.318	1.320
JAE4	1.318	1.328	1.334	1.338	1.339
TOS	1.296±.016	1.305±.016	1.307±.017	1.312±.017	1.314±.018
Average	1.319	1.332	1.336	1.341	1.340

Table 24 : Fission Density AM-Traverse, M-MOX, Voided

Contr.	Pin-Position				
	1.0	2.0	3.0	4.0	5.0
ANL	1.000±.003	1.047±.005	1.161±.006	1.467±.010	2.388±.023
BEN	1.000	1.039	1.137	1.380	2.184
CEA3	1.000	1.057	1.203	1.532	2.316
CEA4	1.000	1.035	1.133	1.398	2.674
CEA5	1.000	1.031	1.120	1.275	2.704
CEN	1.000	1.041	1.156	1.467	2.767
ECN	1.000	1.070±.016	1.172±.019	1.535±.025	2.438±.039
HIT	1.000	1.052	1.167	1.477	2.384
IKE1	1.000	1.043	1.163	1.493	2.654
IKE2	1.000±.006	1.049±.005	1.157±.005	1.461±.007	2.393±.011
JAE1	1.000	1.038	1.159	1.497	2.711
JAE2	1.000±.010	1.046±.007	1.171±.008	1.488±.011	2.453±.019
JAE3	1.000	1.039	1.148	1.443	2.670
JAE4	1.000	1.035	1.140	1.447	2.747
TOS	1.000±.024	1.015±.016	1.109±.017	1.416±.023	2.406±.040
Average	1.000	1.042	1.153	1.452	2.526

Contr.	Pin-Position				
	6.0	7.0	8.0	9.0	10.0
ANL	1.921±.016	2.775±.030	3.280±.043	3.576±.048	3.764±.055
BEN	1.925	2.792	3.275	3.553	3.714
CEA3	1.827	2.533	2.991	3.280	3.458
CEA4	1.890	2.852	3.378	3.707	3.895
CEA5	1.968	2.961	3.495	3.811	3.986
CEN	2.184	3.160	3.707	4.030	4.218
ECN	1.959±.027	2.890±.038	3.341±.043	3.752±.045	3.903±.047
HIT	1.910	2.747	3.230	3.539	3.723
IKE1	2.018	2.898	3.430	3.750	3.940
IKE2	1.888±.008	2.733±.010	3.216±.012	3.519±.012	3.685±.013
JAE1	1.821	2.701	3.250	3.586	3.785
JAE2	1.961±.014	2.861±.017	3.317±.022	3.617±.026	3.850±.023
JAE3	2.031	2.922	3.427	3.701	3.872
JAE4	1.930	2.852	3.415	3.750	3.943
TOS	1.896±.027	2.724±.036	3.274±.043	3.557±.045	3.683±.044
Average	1.942	2.827	3.335	3.648	3.828

Contr.	Pin-Position				
	11.0	12.0	13.0	14.0	15.0
ANL	3.854±.063	3.904±.060	3.940±.062	3.955±.063	3.957±.066
BEN	3.808	3.863	3.894	3.914	3.922
CEA3	3.566	3.629	3.665	3.684	3.693
CEA4	4.004	4.064	4.096	4.112	4.118
CEA5	4.085	4.141	4.170	4.186	4.191
CEN	4.328	4.390	4.425	4.444	4.452
ECN	3.996±.048	3.965±.048	4.085±.049	4.118±.049	4.059±.053
HIT	3.822	3.871	3.877	3.909	3.911
IKE1	4.049	4.110	4.143	4.160	4.166
IKE2	3.806±.013	3.871±.013	3.883±.013	3.911±.013	3.911±.014
JAE1	3.901	3.966	4.001	4.018	4.026
JAE2	3.870±.022	3.948±.023	3.956±.024	4.003±.023	4.012±.030
JAE3	3.975	4.036	4.070	4.089	4.097
JAE4	4.052	4.111	4.141	4.156	4.162
TOS	3.715±.045	3.816±.045	3.867±.046	3.921±.047	4.033±.053
Average	3.922	3.979	4.014	4.039	4.047

Table 25 : Fission Density AD-Traverse, M-MOX, Moderated

Contr.	Pin-Position				
	1.0	2.0	3.0	4.0	5.0
ANL	1.000±.004	1.013±.006	1.071±.007	1.239±.008	1.899±.021
BEN	1.000	1.018	1.074	1.246	1.902
CEA3	1.000	1.027	1.106	1.315	1.874
CEA4	1.000	1.010	1.072	1.173	2.007
CEA5	1.000	1.011	1.069	1.164	1.986
CEN	1.000	1.018	1.075	1.252	1.890
ECN	1.000	1.032±.017	1.066±.017	1.225±.018	1.887±.025
HIT	1.000	1.018	1.072	1.235	1.863
IKE1	1.000	1.019	1.079	1.266	1.895
IKE2	1.000±.007	1.019±.007	1.064±.007	1.250±.008	1.904±.010
JAE1	1.000	1.016	1.072	1.261	1.931
JAE2	1.000±.011	1.004±.010	1.059±.012	1.230±.012	1.875±.016
JAE3	1.000	1.019	1.075	1.256	1.868
JAE4	1.000	1.017	1.073	1.259	1.926
TOS	1.000±.026	0.961±.023	1.033±.024	1.211±.026	1.871±.035
Average	1.000	1.013	1.071	1.239	1.905

Contr.	Pin-Position				
	6.0	7.0	8.0	9.0	10.0
ANL	1.173±.007	1.294±.009	1.337±.009	1.359±.009	1.372±.009
BEN	1.164	1.303	1.355	1.375	1.386
CEA3	1.128	1.246	1.297	1.320	1.332
CEA4	1.150	1.287	1.340	1.362	1.373
CEA5	1.143	1.280	1.332	1.354	1.364
CEN	1.136	1.260	1.306	1.325	1.335
ECN	1.139±.014	1.308±.016	1.359±.015	1.355±.015	1.384±.015
HIT	1.136	1.262	1.299	1.318	1.324
IKE1	1.138	1.267	1.315	1.336	1.346
IKE2	1.124±.006	1.241±.006	1.289±.006	1.317±.006	1.319±.006
JAE1	1.128	1.256	1.306	1.327	1.338
JAE2	1.124±.009	1.252±.010	1.307±.009	1.331±.010	1.347±.011
JAE3	1.124	1.250	1.293	1.313	1.323
JAE4	1.127	1.262	1.312	1.333	1.343
TOS	1.081±.019	1.239±.021	1.301±.021	1.297±.021	1.285±.020
Average	1.134	1.267	1.316	1.335	1.345

Contr.	Pin-Position				
	11.0	12.0	13.0	14.0	15.0
ANL	1.369±.009	1.385±.009	1.394±.010	1.385±.009	1.384±.011
BEN	1.393	1.397	1.402	1.403	1.404
CEA3	1.339	1.343	1.345	1.347	1.348
CEA4	1.380	1.384	1.387	1.389	1.390
CEA5	1.371	1.374	1.377	1.378	1.379
CEN	1.341	1.345	1.347	1.349	1.349
ECN	1.377±.015	1.378±.015	1.385±.015	1.390±.017	1.368±.019
HIT	1.332	1.338	1.341	1.345	1.349
IKE1	1.352	1.356	1.358	1.360	1.360
IKE2	1.335±.006	1.336±.006	1.339±.006	1.348±.006	1.346±.007
JAE1	1.344	1.347	1.350	1.351	1.352
JAE2	1.344±.010	1.351±.010	1.341±.009	1.343±.011	1.337±.010
JAE3	1.329	1.333	1.335	1.336	1.337
JAE4	1.348	1.352	1.354	1.356	1.356
TOS	1.378±.023	1.333±.022	1.359±.023	1.304±.022	1.321±.026
Average	1.355	1.357	1.361	1.359	1.359

Table 26 : Fission Density AD-Traverse, M-MOX, Voided

Contr.	Pin-Position				
	1.0	2.0	3.0	4.0	5.0
ANL	1.000±.003	1.092±.006	1.326±.009	1.962±.017	3.765±.065
BEN	1.000	1.083	1.285	1.794	3.455
CEA3	1.000	1.112	1.395	2.037	3.651
CEA4	1.000	1.079	1.278	1.787	4.605
CEA5	1.000	1.073	1.252	1.726	4.679
CEN	1.000	1.083	1.313	1.950	4.688
ECN	1.000	1.113±.017	1.371±.021	1.949±.031	3.843±.054
HIT	1.000	1.098	1.338	1.946	3.759
IKE1	1.000	1.085	1.329	2.009	4.492
IKE2	1.000±.006	1.086±.006	1.321±.008	1.954±.012	3.781±.021
JAE1	1.000	1.080	1.324	2.029	4.674
JAE2	1.000±.010	1.085±.010	1.316±.013	2.021±.020	3.810±.035
JAE3	1.000	1.078	1.294	1.893	4.489
JAE4	1.000	1.071	1.281	1.919	4.753
TOS	1.000±.024	1.061±.022	1.302±.028	1.998±.045	3.770±.074
Average	1.000	1.085	1.315	1.932	4.148

Contr.	Pin-Position				
	6.0	7.0	8.0	9.0	10.0
ANL	2.912±.039	3.538±.052	3.810±.066	3.910±.068	3.953±.072
BEN	2.881	3.542	3.787	3.877	3.917
CEA3	2.725	3.273	3.518	3.631	3.682
CEA4	2.914	3.649	3.942	4.061	4.110
CEA5	3.023	3.749	4.032	4.145	4.190
CEN	3.287	4.012	4.282	4.397	4.448
ECN	2.983±.036	3.627±.040	3.892±.043	4.017±.044	4.066±.045
HIT	2.900	3.524	3.738	3.880	3.902
IKE1	3.065	3.743	4.009	4.117	4.160
IKE2	2.882±.014	3.484±.016	3.726±.016	3.849±.017	3.888±.017
JAE1	2.848	3.564	3.849	3.968	4.017
JAE2	2.961±.026	3.603±.027	3.893±.026	3.991±.029	4.008±.031
JAE3	3.032	3.701	3.939	4.041	4.088
JAE4	2.996	3.722	4.000	4.111	4.154
TOS	2.906±.050	3.477±.057	3.749±.059	3.808±.058	3.900±.060
Average	2.954	3.614	3.878	3.987	4.032

Contr.	Pin-Position				
	11.0	12.0	13.0	14.0	15.0
ANL	3.985±.068	3.995±.067	3.997±.066	3.999±.077	4.004±.064
BEN	3.937	3.945	3.949	3.953	3.953
CEA3	3.706	3.717	3.722	3.725	3.726
CEA4	4.132	4.141	4.144	4.146	4.147
CEA5	4.210	4.218	4.220	4.222	4.223
CEN	4.473	4.486	4.492	4.496	4.498
ECN	4.122±.045	4.122±.045	4.066±.045	4.097±.045	4.097±.053
HIT	3.930	3.925	3.920	3.947	3.941
IKE1	4.177	4.184	4.187	4.188	4.189
IKE2	3.893±.017	3.939±.017	3.941±.017	3.946±.017	3.912±.021
JAE1	4.038	4.046	4.050	4.052	4.053
JAE2	4.055±.031	4.043±.030	4.042±.027	4.002±.034	3.947±.037
JAE3	4.110	4.120	4.124	4.126	4.127
JAE4	4.171	4.177	4.180	4.181	4.182
TOS	3.954±.062	3.880±.059	3.817±.061	3.899±.061	3.959±.074
Average	4.059	4.062	4.057	4.065	4.064

Table 27 : Fission Density AM-Traverse, L-MOX, Moderated

Contr.	Pin-Position				
	1.0	2.0	3.0	4.0	5.0
ANL	1.000±.003	1.001±.004	1.034±.004	1.115±.005	1.359±.007
BEN	1.000	1.010	1.037	1.111	1.357
CEA3	1.000	1.013	1.048	1.135	1.348
CEA4	1.000	1.007	1.037	1.102	1.398
CEA5	1.000	1.007	1.035	1.098	1.393
CEN	1.000	1.009	1.037	1.112	1.349
ECN	1.000	0.983±.017	1.039±.017	1.088±.016	1.366±.020
HIT	1.000	1.015	1.036	1.120	1.353
IKE1	1.000	1.010	1.039	1.119	1.354
IKE2	1.000±.007	1.020±.005	1.044±.005	1.117±.005	1.369±.006
JAE1	1.000	1.008	1.036	1.119	1.373
JAE2	1.000±.012	1.019±.007	1.042±.007	1.120±.008	1.350±.009
JAE3	1.000	1.010	1.036	1.113	1.347
JAE4	1.000	1.009	1.037	1.118	1.368
TOS	1.000±.026	0.978±.017	0.979±.016	1.075±.017	1.323±.020
Average	1.000	1.007	1.034	1.111	1.360

Contr.	Pin-Position				
	6.0	7.0	8.0	9.0	10.0
ANL	1.057±.005	1.217±.006	1.306±.007	1.346±.007	1.378±.007
BEN	1.056	1.228	1.317	1.364	1.389
CEA3	1.027	1.170	1.256	1.308	1.338
CEA4	1.036	1.216	1.306	1.357	1.385
CEA5	1.034	1.216	1.307	1.356	1.383
CEN	1.026	1.185	1.267	1.312	1.337
ECN	1.078±.015	1.224±.016	1.321±.017	1.372±.018	1.405±.018
HIT	1.039	1.200	1.287	1.334	1.358
IKE1	1.031	1.191	1.280	1.328	1.355
IKE2	1.030±.004	1.183±.004	1.265±.005	1.315±.005	1.343±.005
JAE1	1.026	1.183	1.272	1.322	1.350
JAE2	1.059±.007	1.219±.008	1.308±.009	1.347±.010	1.389±.008
JAE3	1.025	1.186	1.271	1.313	1.339
JAE4	1.022	1.188	1.280	1.330	1.357
TOS	0.986±.014	1.148±.015	1.211±.016	1.258±.016	1.286±.016
Average	1.036	1.197	1.284	1.331	1.360

Contr.	Pin-Position				
	11.0	12.0	13.0	14.0	15.0
ANL	1.393±.008	1.403±.008	1.409±.007	1.411±.008	1.414±.007
BEN	1.405	1.415	1.421	1.425	1.427
CEA3	1.356	1.367	1.373	1.377	1.378
CEA4	1.401	1.410	1.416	1.419	1.421
CEA5	1.398	1.407	1.413	1.416	1.417
CEN	1.352	1.360	1.365	1.368	1.370
ECN	1.389±.018	1.383±.017	1.383±.017	1.395±.018	1.398±.020
HIT	1.370	1.377	1.380	1.386	1.388
IKE1	1.371	1.380	1.385	1.388	1.390
IKE2	1.364±.005	1.366±.005	1.368±.005	1.371±.005	1.383±.005
JAE1	1.366	1.376	1.381	1.384	1.386
JAE2	1.378±.008	1.400±.010	1.400±.008	1.395±.007	1.409±.009
JAE3	1.355	1.364	1.370	1.374	1.375
JAE4	1.372	1.381	1.386	1.389	1.390
TOS	1.293±.016	1.319±.017	1.321±.017	1.318±.017	1.296±.018
Average	1.371	1.381	1.385	1.388	1.389

Table 28 : Fission Density AM-Traverse, L-MOX, Voided

Contr.	Pin-Position				
	1.0	2.0	3.0	4.0	5.0
ANL	1.000±.006	1.046±.007	1.159±.008	1.437±.013	2.149±.031
BEN	1.000	1.039	1.130	1.357	1.983
CEA3	1.000	1.057	1.196	1.479	2.086
CEA4	1.000	1.046	1.154	1.423	2.389
CEA5	1.000	1.043	1.148	1.412	2.445
CEN	1.000	1.051	1.181	1.483	2.418
ECN	1.000	1.076±.017	1.200±.019	1.476±.024	2.218±.035
HIT	1.000	1.056	1.168	1.440	2.152
IKE1	1.000	1.056	1.197	1.518	2.368
IKE2	1.000±.007	1.055±.005	1.183±.005	1.467±.007	2.208±.010
JAE1	1.000	1.051	1.195	1.532	2.452
JAE2	1.000±.011	1.028±.008	1.160±.009	1.415±.011	2.123±.018
JAE3	1.000	1.049	1.175	1.471	2.404
JAE4	1.000	1.053	1.191	1.521	2.524
TOS	1.000±.027	1.036±.019	1.121±.018	1.387±.023	2.095±.034
Average	1.000	1.049	1.171	1.454	2.268

Contr.	Pin-Position				
	6.0	7.0	8.0	9.0	10.0
ANL	2.252±.032	3.178±.064	3.723±.084	4.024±.096	4.208±.106
BEN	2.257	3.187	3.691	3.974	4.139
CEA3	2.123	2.877	3.354	3.650	3.830
CEA4	2.262	3.319	3.874	4.215	4.404
CEA5	2.357	3.464	4.037	4.371	4.552
CEN	2.512	3.554	4.123	4.452	4.641
ECN	2.320±.032	3.246±.042	3.870±.046	4.094±.049	4.298±.052
HIT	2.257	3.159	3.695	4.004	4.173
IKE1	2.366	3.332	3.897	4.230	4.422
IKE2	2.252±.009	3.204±.012	3.713±.013	4.042±.014	4.206±.014
JAE1	2.205	3.176	3.764	4.114	4.317
JAE2	2.220±.015	3.137±.021	3.656±.024	3.998±.024	4.105±.028
JAE3	2.407	3.387	3.927	4.214	4.391
JAE4	2.347	3.391	4.008	4.365	4.567
TOS	2.106±.030	3.024±.041	3.587±.045	3.840±.046	4.019±.048
Average	2.283	3.242	3.795	4.106	4.285

Contr.	Pin-Position				
	11.0	12.0	13.0	14.0	15.0
ANL	4.319±.113	4.363±.113	4.388±.117	4.412±.123	4.420±.117
BEN	4.235	4.291	4.322	4.339	4.348
CEA3	3.937	3.999	4.034	4.053	4.061
CEA4	4.512	4.571	4.602	4.617	4.623
CEA5	4.655	4.711	4.742	4.757	4.763
CEN	4.750	4.811	4.846	4.864	4.872
ECN	4.400±.053	4.391±.053	4.369±.052	4.518±.054	4.478±.058
HIT	4.269	4.318	4.366	4.359	4.380
IKE1	4.531	4.591	4.623	4.639	4.645
IKE2	4.304±.014	4.349±.014	4.370±.014	4.404±.015	4.398±.016
JAE1	4.433	4.497	4.532	4.549	4.556
JAE2	4.187±.024	4.265±.024	4.321±.026	4.279±.027	4.296±.029
JAE3	4.496	4.558	4.592	4.611	4.620
JAE4	4.678	4.737	4.768	4.782	4.788
TOS	4.222±.051	4.128±.050	4.212±.050	4.229±.052	4.144±.055
Average	4.395	4.439	4.472	4.494	4.493

Table 29 : Fission Density AD-Traverse, L-MOX, Moderated

Contr.	Pin-Position				
	1.0	2.0	3.0	4.0	5.0
ANL	1.000±.003	1.019±.004	1.066±.006	1.213±.007	1.644±.012
BEN	1.000	1.019	1.074	1.215	1.634
CEA3	1.000	1.025	1.094	1.252	1.613
CEA4	1.000	1.014	1.072	1.184	1.703
CEA5	1.000	1.015	1.069	1.176	1.693
CEN	1.000	1.018	1.072	1.212	1.617
ECN	1.000	1.018±.016	1.072±.016	1.202±.018	1.656±.022
HIT	1.000	1.024	1.071	1.220	1.627
IKE1	1.000	1.019	1.075	1.226	1.632
IKE2	1.000±.007	1.029±.007	1.082±.007	1.214±.007	1.650±.009
JAE1	1.000	1.017	1.072	1.228	1.661
JAE2	1.000±.012	1.037±.011	1.073±.010	1.226±.012	1.629±.013
JAE3	1.000	1.019	1.070	1.214	1.615
JAE4	1.000	1.018	1.072	1.224	1.657
TOS	1.000±.026	1.008±.023	1.014±.022	1.156±.024	1.498±.028
Average	1.000	1.020	1.070	1.211	1.635

Contr.	Pin-Position				
	6.0	7.0	8.0	9.0	10.0
ANL	1.252±.007	1.357±.009	1.398±.009	1.408±.009	1.416±.008
BEN	1.254	1.366	1.406	1.424	1.434
CEA3	1.216	1.311	1.352	1.371	1.382
CEA4	1.243	1.354	1.397	1.416	1.426
CEA5	1.241	1.351	1.394	1.413	1.422
CEN	1.215	1.313	1.349	1.365	1.374
ECN	1.249±.015	1.360±.016	1.389±.015	1.454±.016	1.415±.016
HIT	1.231	1.333	1.375	1.385	1.398
IKE1	1.227	1.330	1.369	1.386	1.395
IKE2	1.218±.006	1.327±.006	1.354±.006	1.372±.006	1.385±.006
JAE1	1.219	1.322	1.363	1.381	1.390
JAE2	1.243±.009	1.352±.012	1.388±.011	1.412±.010	1.404±.011
JAE3	1.215	1.317	1.353	1.370	1.379
JAE4	1.220	1.328	1.369	1.386	1.395
TOS	1.160±.020	1.266±.021	1.310±.021	1.320±.021	1.317±.022
Average	1.227	1.332	1.371	1.391	1.395

Contr.	Pin-Position				
	11.0	12.0	13.0	14.0	15.0
ANL	1.428±.008	1.429±.009	1.429±.009	1.438±.009	1.431±.010
BEN	1.440	1.446	1.448	1.450	1.451
CEA3	1.388	1.392	1.394	1.396	1.397
CEA4	1.432	1.436	1.439	1.441	1.442
CEA5	1.428	1.432	1.435	1.436	1.437
CEN	1.379	1.383	1.385	1.387	1.387
ECN	1.416±.016	1.425±.016	1.440±.016	1.406±.017	1.424±.020
HIT	1.397	1.399	1.407	1.412	1.410
IKE1	1.400	1.404	1.407	1.408	1.409
IKE2	1.393±.006	1.385±.006	1.392±.006	1.399±.007	1.394±.008
JAE1	1.396	1.400	1.402	1.404	1.405
JAE2	1.423±.011	1.438±.011	1.425±.011	1.446±.010	1.448±.013
JAE3	1.384	1.388	1.390	1.392	1.392
JAE4	1.400	1.404	1.406	1.408	1.409
TOS	1.305±.022	1.336±.022	1.358±.023	1.333±.022	1.367±.027
Average	1.401	1.406	1.410	1.410	1.414

Table 30 : Fission Density AD-Traverse, L-MOX, Voided

Contr.	Pin-Position				
	1.0	2.0	3.0	4.0	5.0
ANL	1.000±.006	1.092±.008	1.341±.013	1.882±.023	3.290±.071
BEN	1.000	1.083	1.272	1.732	3.004
CEA3	1.000	1.113	1.383	1.937	3.161
CEA4	1.000	1.092	1.317	1.838	3.894
CEA5	1.000	1.089	1.307	1.820	4.015
CEN	1.000	1.103	1.363	1.972	3.875
ECN	1.000	1.102±.017	1.366±.020	1.896±.028	3.324±.047
HIT	1.000	1.095	1.351	1.926	3.272
IKE1	1.000	1.112	1.395	2.043	3.777
IKE2	1.000±.007	1.102±.007	1.355±.008	1.927±.012	3.373±.019
JAE1	1.000	1.107	1.396	2.081	3.973
JAE2	1.000±.011	1.103±.011	1.329±.014	1.884±.017	3.300±.030
JAE3	1.000	1.099	1.349	1.943	3.849
JAE4	1.000	1.105	1.382	2.054	4.123
TOS	1.000±.027	1.054±.023	1.318±.028	1.841±.039	3.202±.063
Average	1.000	1.097	1.348	1.918	3.562

Contr.	Pin-Position				
	6.0	7.0	8.0	9.0	10.0
ANL	3.334±.074	3.996±.100	4.248±.114	4.365±.118	4.421±.124
BEN	3.272	3.965	4.211	4.307	4.346
CEA3	3.073	3.641	3.890	4.002	4.053
CEA4	3.381	4.150	4.449	4.569	4.618
CEA5	3.525	4.301	4.598	4.715	4.762
CEN	3.682	4.432	4.705	4.820	4.872
ECN	3.422±.041	4.060±.045	4.282±.047	4.521±.050	4.563±.050
HIT	3.320	3.981	4.197	4.326	4.358
IKE1	3.512	4.222	4.492	4.600	4.643
IKE2	3.316±.016	4.022±.018	4.258±.019	4.373±.019	4.419±.019
JAE1	3.340	4.092	4.383	4.502	4.551
JAE2	3.289±.025	3.921±.028	4.180±.036	4.240±.035	4.319±.033
JAE3	3.503	4.214	4.460	4.565	4.614
JAE4	3.556	4.336	4.626	4.739	4.783
TOS	3.168±.055	3.823±.061	4.017±.064	4.181±.066	4.308±.068
Average	3.380	4.077	4.333	4.455	4.509

Contr.	Pin-Position				
	11.0	12.0	13.0	14.0	15.0
ANL	4.452±.123	4.442±.126	4.450±.127	4.473±.120	4.476±.127
BEN	4.368	4.377	4.382	4.386	4.386
CEA3	4.076	4.087	4.093	4.096	4.097
CEA4	4.640	4.649	4.654	4.656	4.658
CEA5	4.783	4.792	4.795	4.797	4.799
CEN	4.898	4.911	4.919	4.924	4.925
ECN	4.470±.049	4.604±.051	4.577±.050	4.549±.050	4.456±.058
HIT	4.396	4.392	4.417	4.418	4.419
IKE1	4.660	4.668	4.672	4.674	4.675
IKE2	4.453±.019	4.449±.019	4.438±.019	4.433±.020	4.404±.023
JAE1	4.572	4.582	4.586	4.589	4.590
JAE2	4.343±.031	4.380±.033	4.374±.034	4.371±.040	4.365±.039
JAE3	4.636	4.647	4.652	4.655	4.656
JAE4	4.800	4.808	4.811	4.813	4.814
TOS	4.187±.064	4.167±.063	4.161±.066	4.265±.067	4.113±.076
Average	4.516	4.530	4.532	4.540	4.522

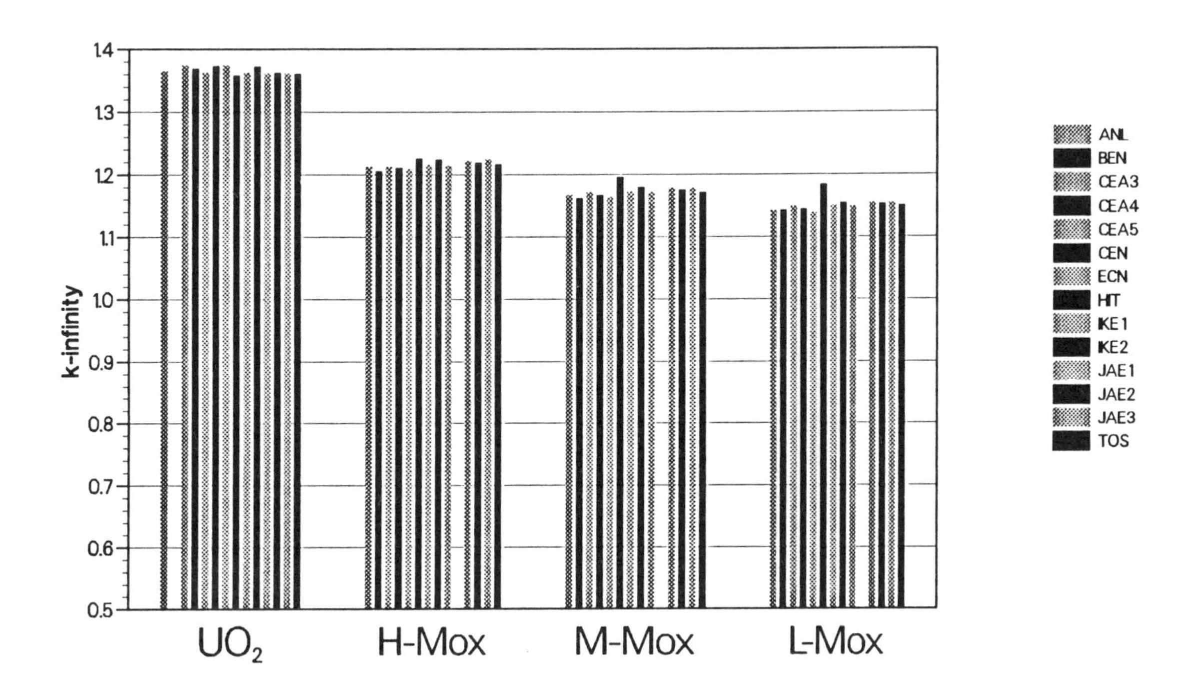

Fig. 1: k-infinity, Pincell, Moderated

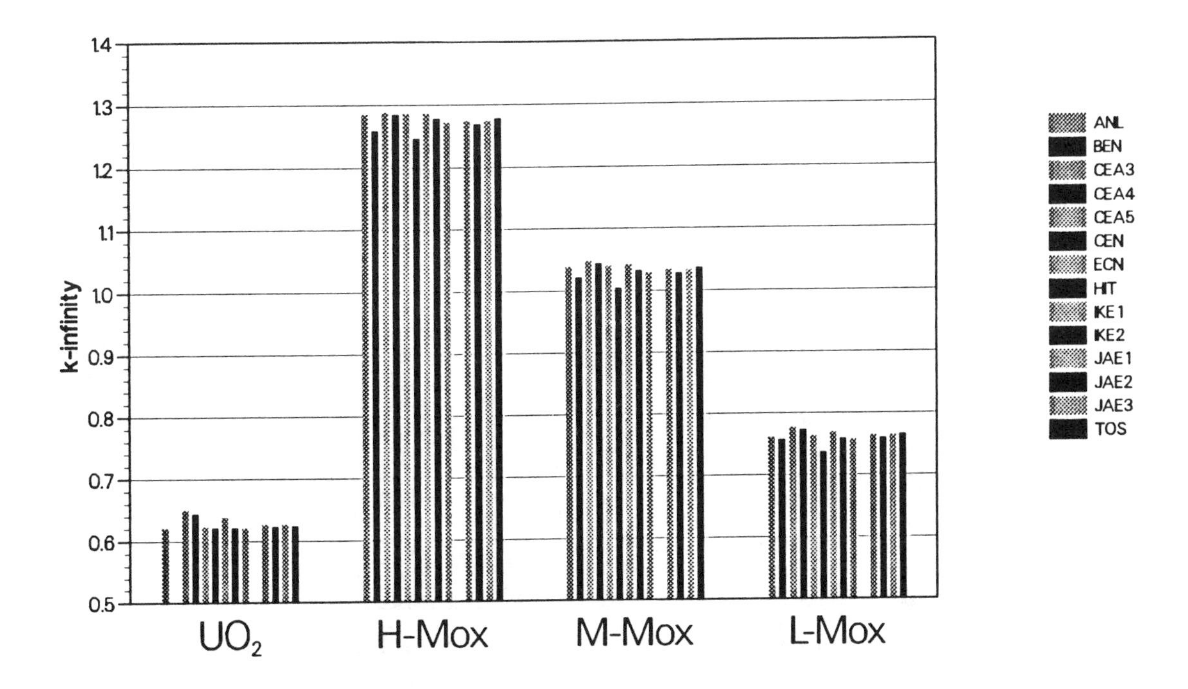

Fig. 2: k-infinity, Pincell, Voided

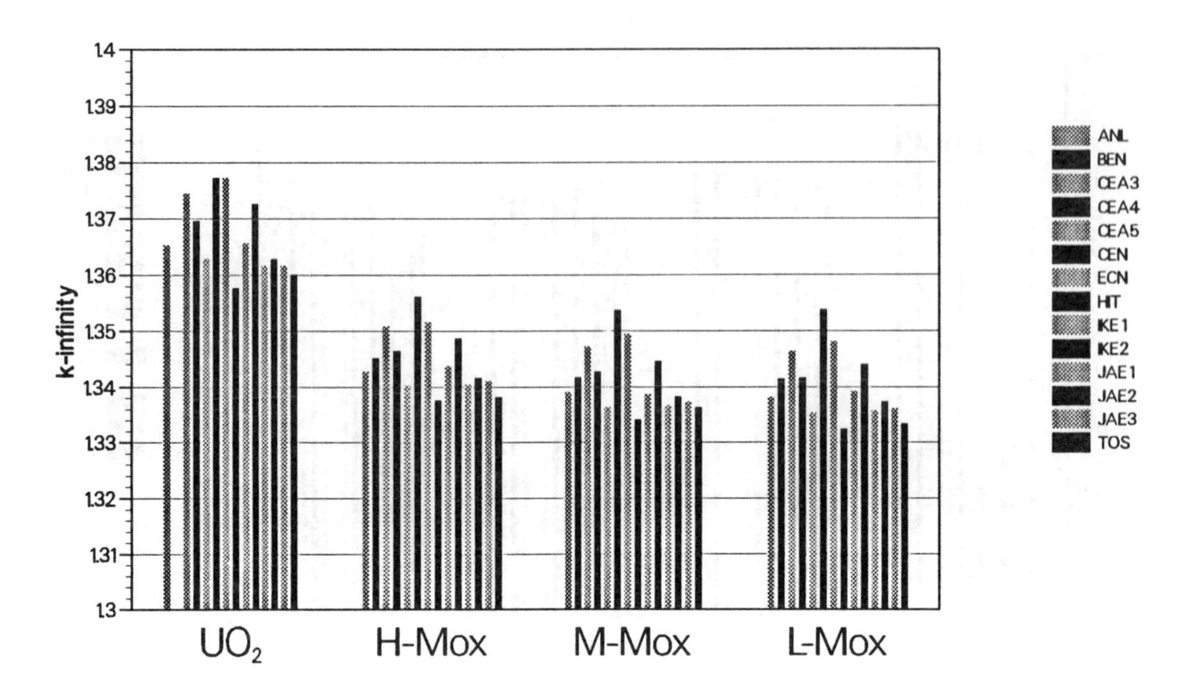

Fig. 3: k-infinity, Macrocell, Moderated

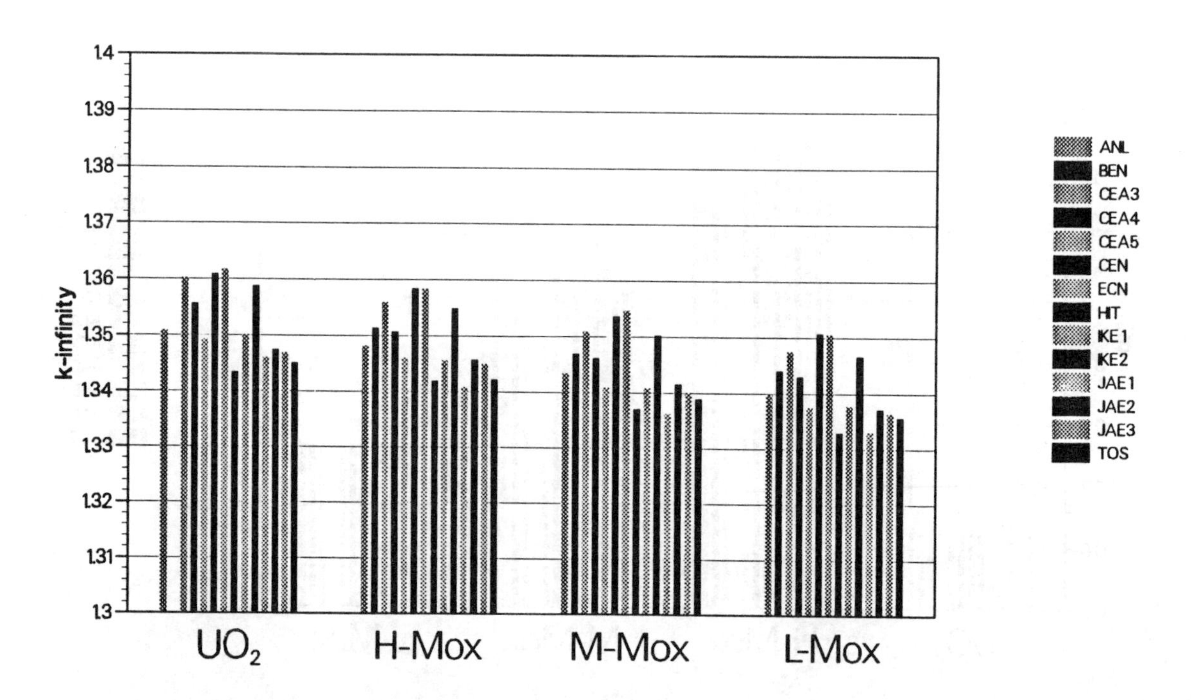

Fig. 4: k-infinity, Macrocell, Voided

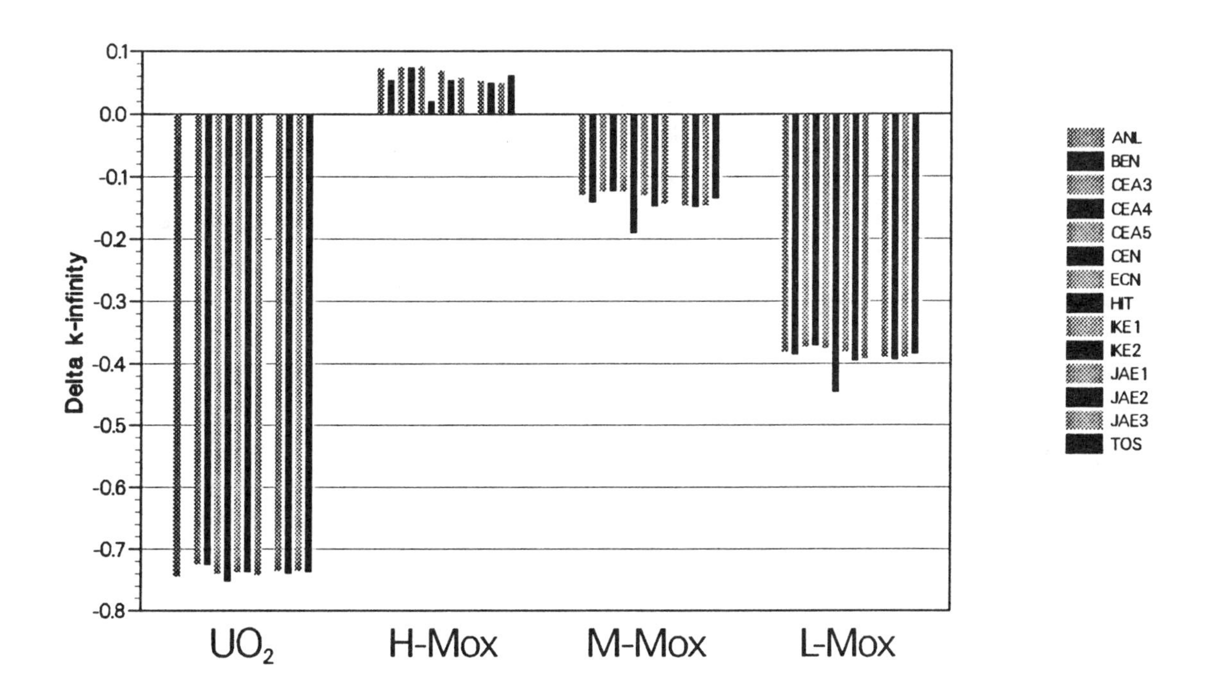

Fig. 5: Delta-k-Void, Pincell

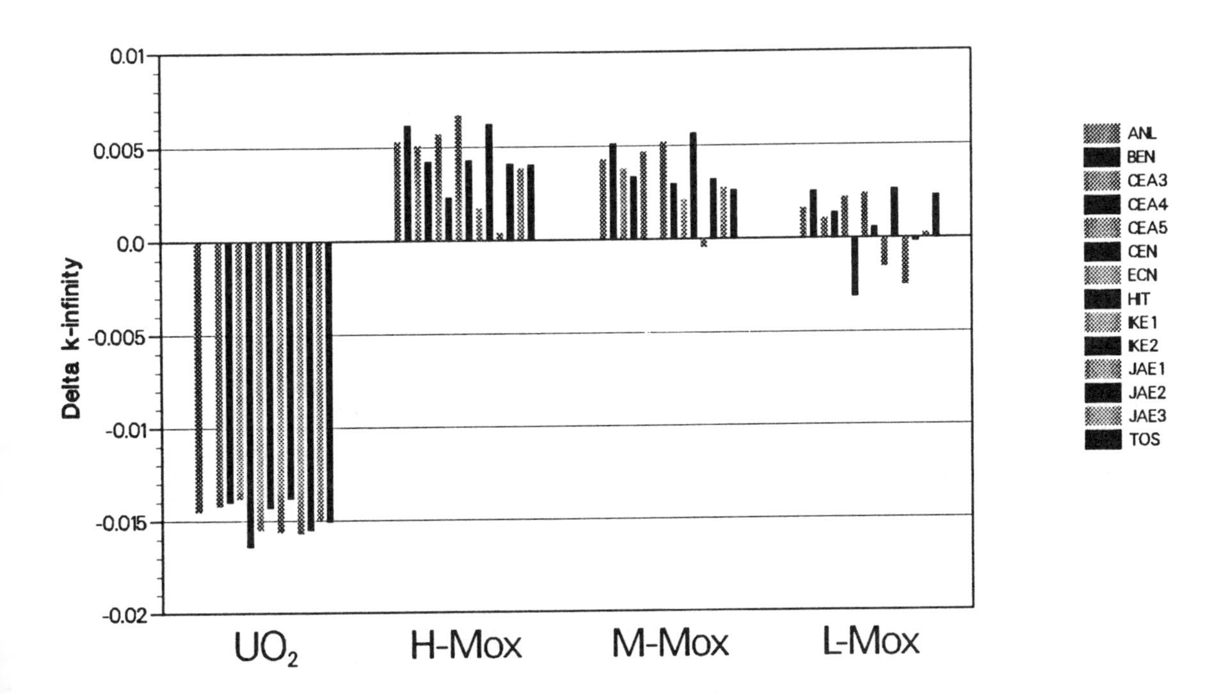

Fig. 6: Delta-k-Void, Macrocell

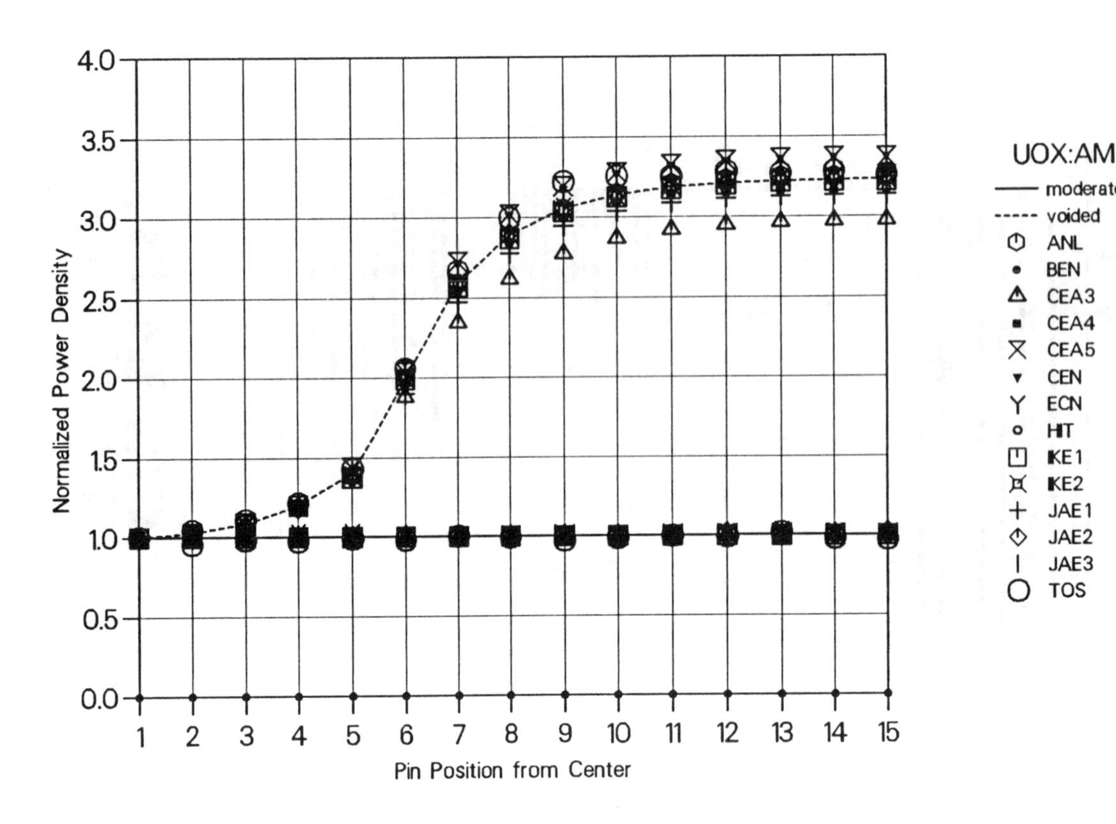

Fig. 7: Fission Density, Traverse AM, UOX

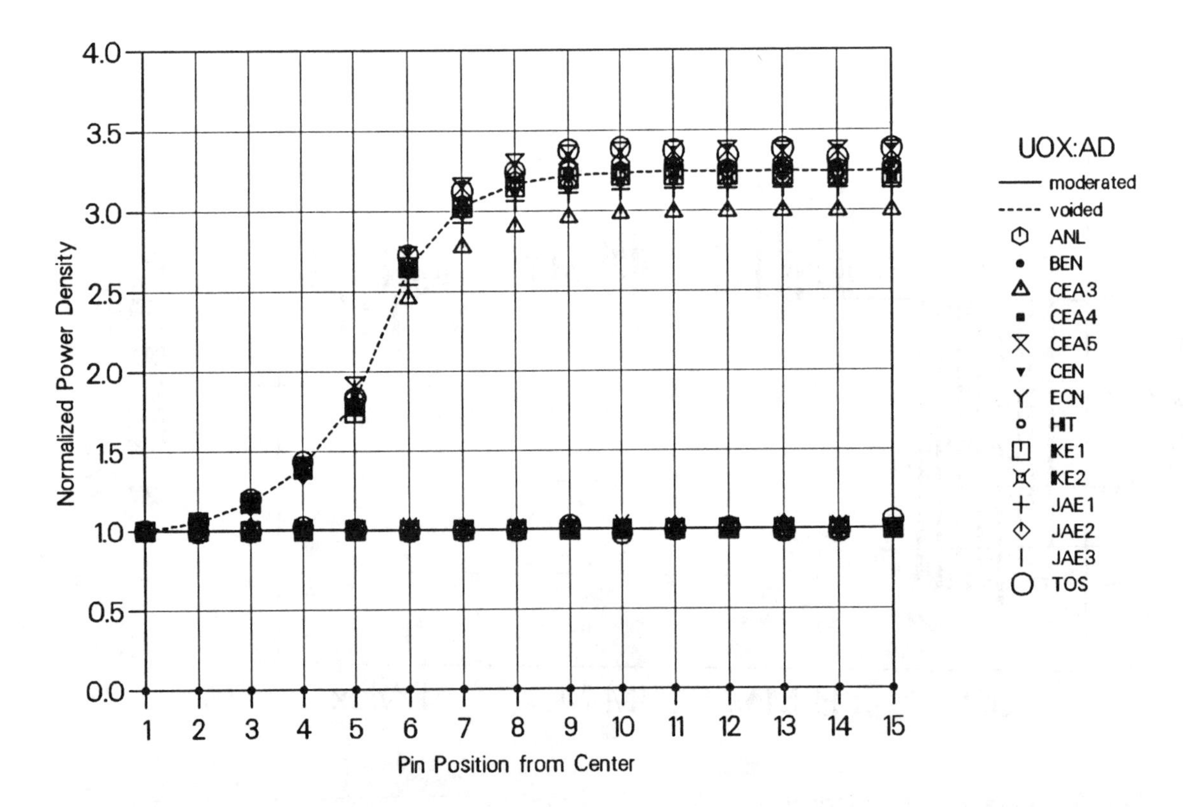

Fig. 8: Fission Density, Traverse AD, UOX

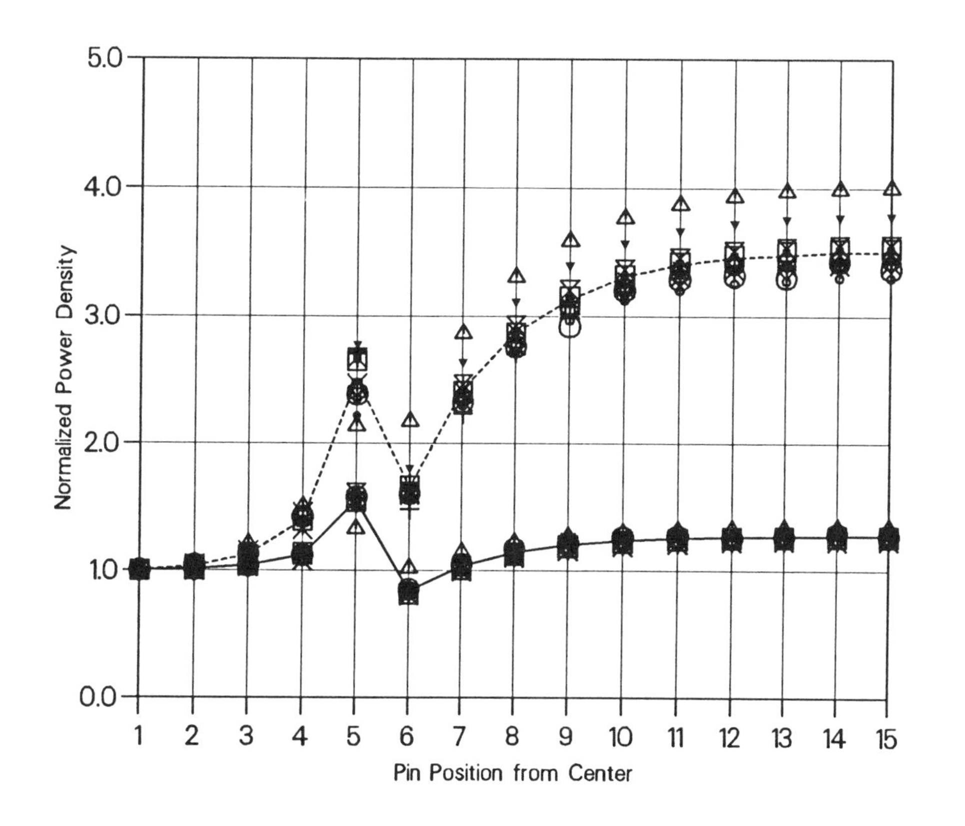

H-MOX:AM
—— moderated
----- voided
ANL
BEN
CEA3
CEA4
CEA5
CEN
ECN
HIT
IKE1
IKE2
JAE1
JAE2
JAE3
TOS

Fig. 9: Fission Density, Traverse AM, H–MOX

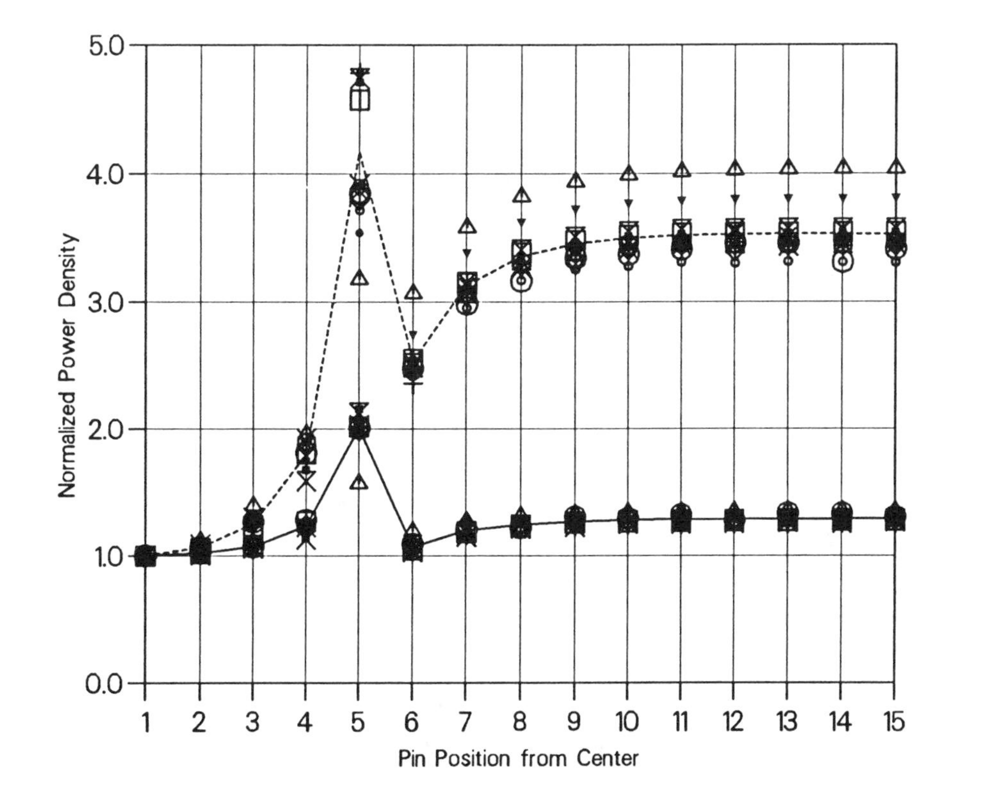

H-MOX:AD
—— moderated
----- voided
ANL
BEN
CEA3
CEA4
CEA5
CEN
ECN
HIT
IKE1
IKE2
JAE1
JAE2
JAE3
TOS

Fig. 10: Fission Density, Traverse AD, H–MOX

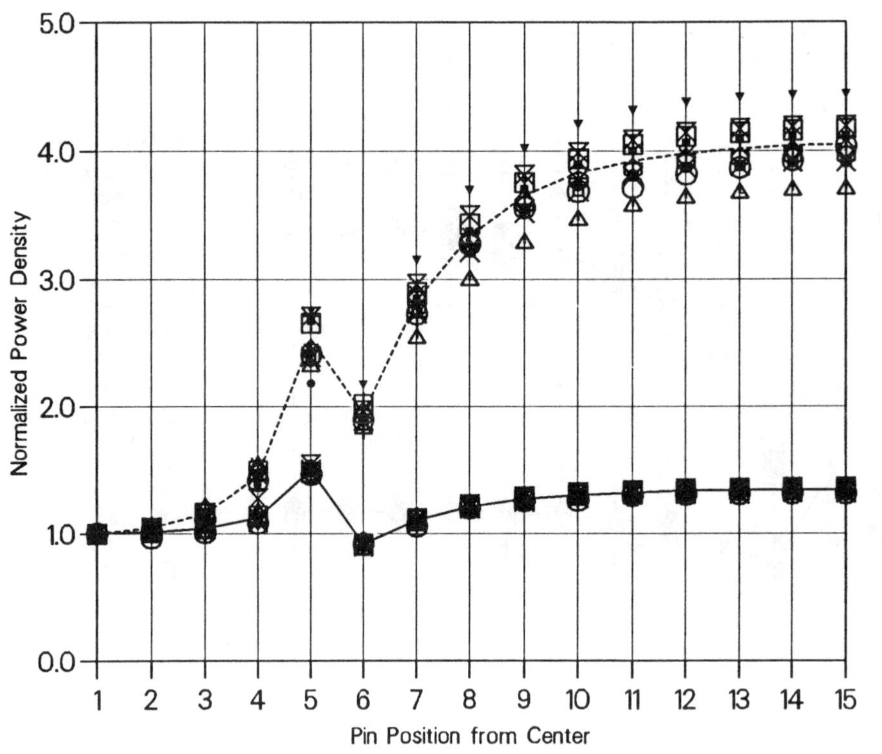

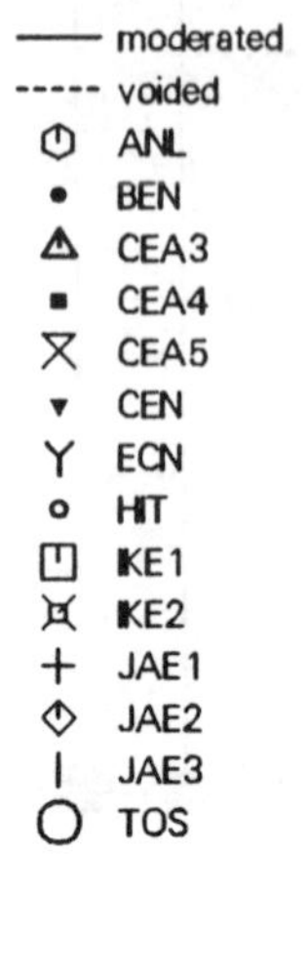

Fig. 11: Fission Density, Traverse AM, M–MOX

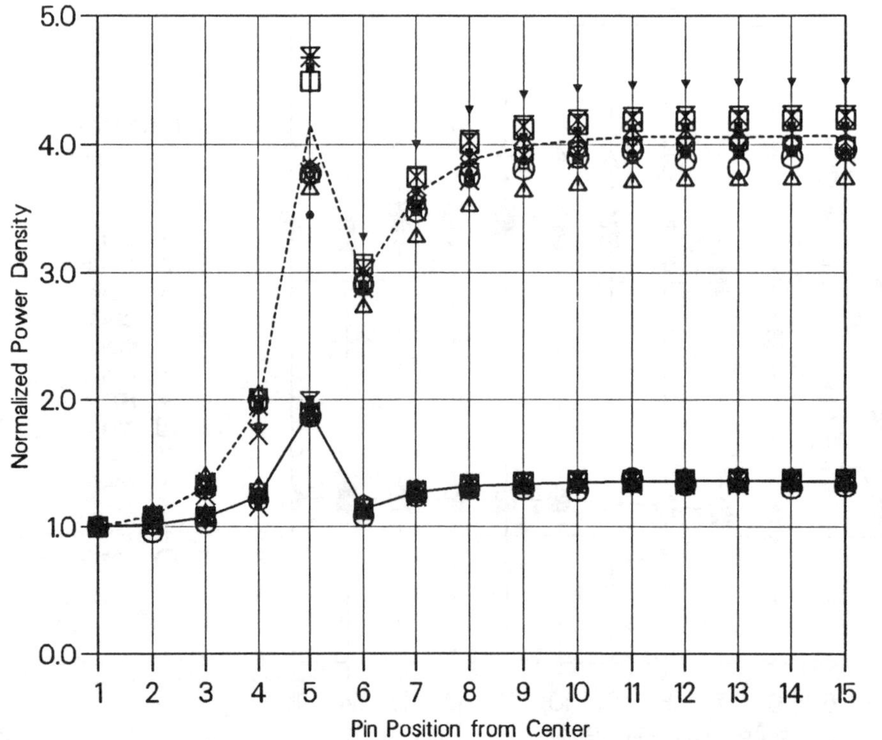

Fig. 12: Fission Density, Traverse AD, M–MOX

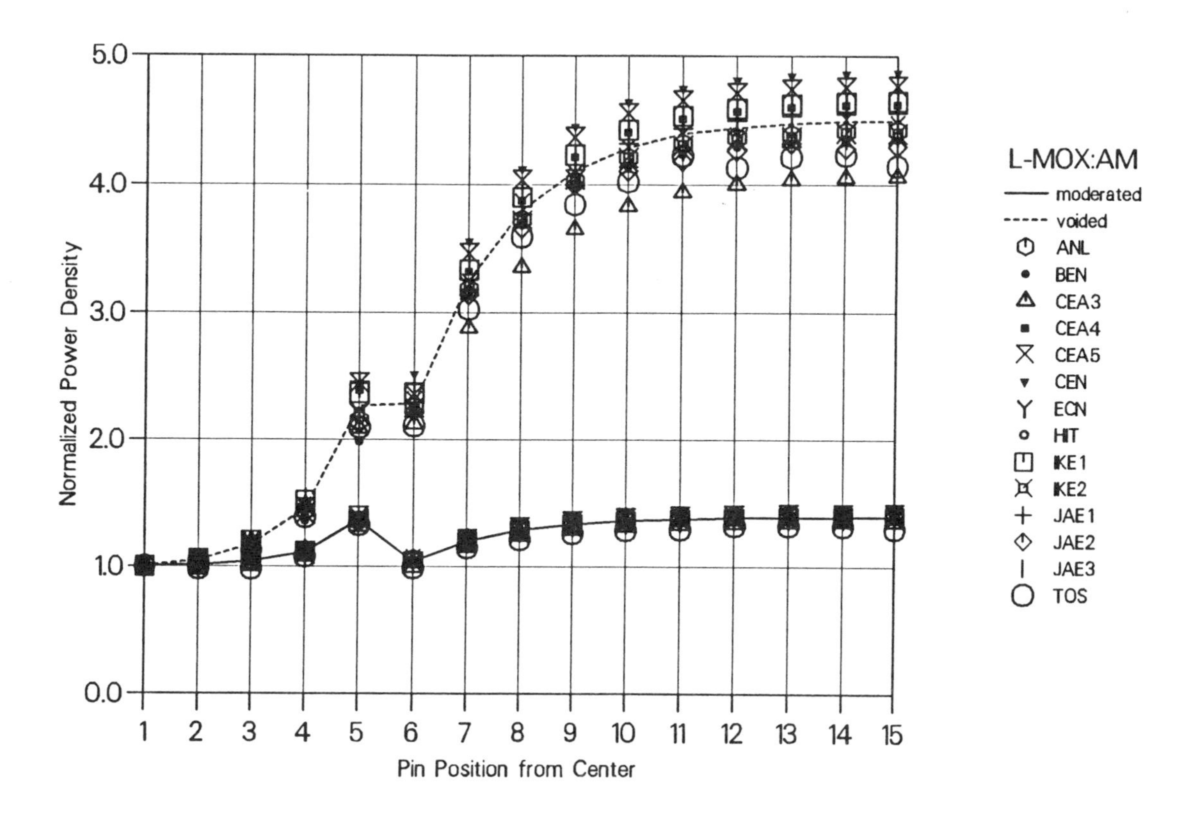

Fig. 13: Fission Density, Traverse AM, L-MOX

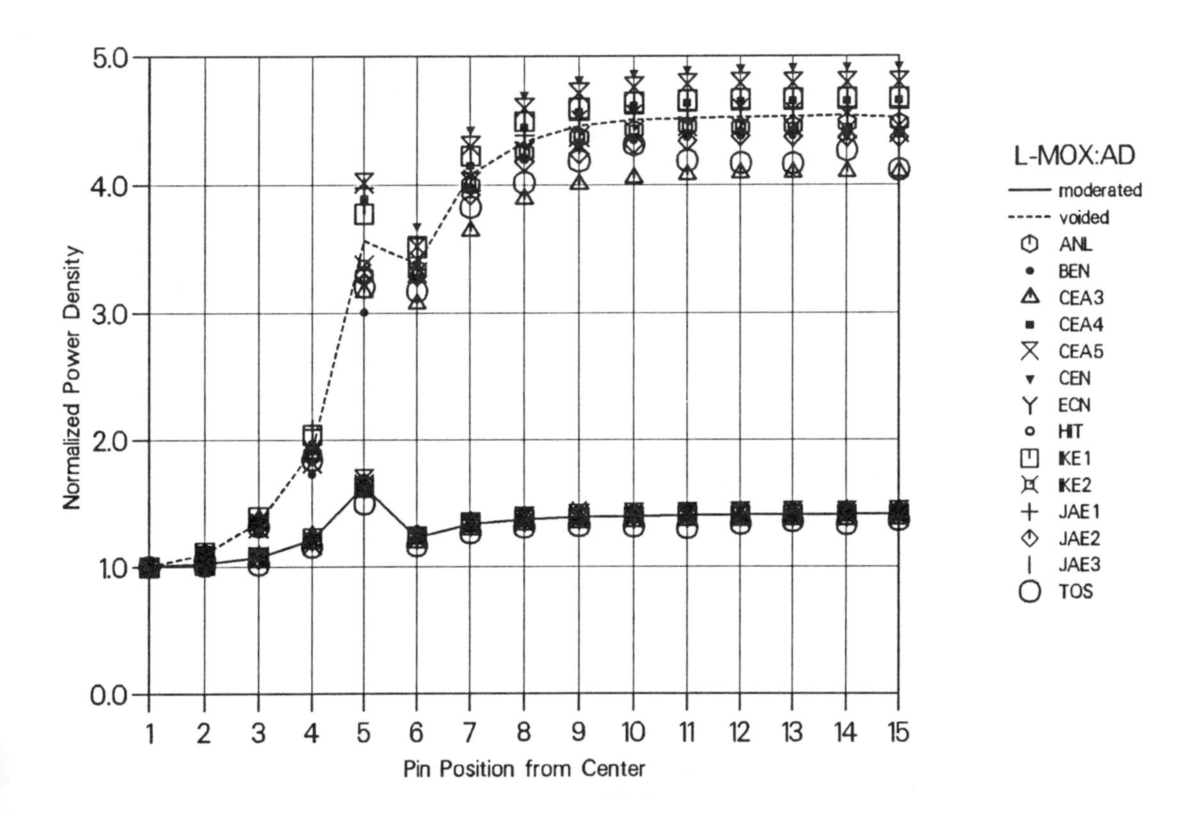

Fig. 14: Fission Density, Traverse AD, L-MOX

H–Mox: ANL

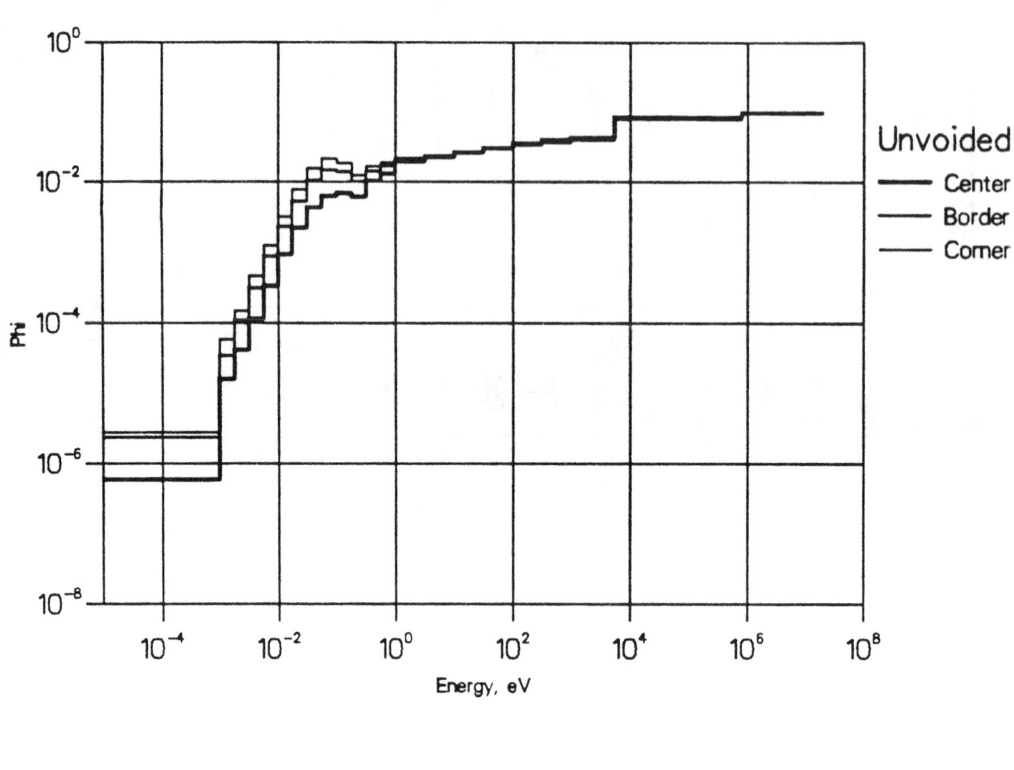

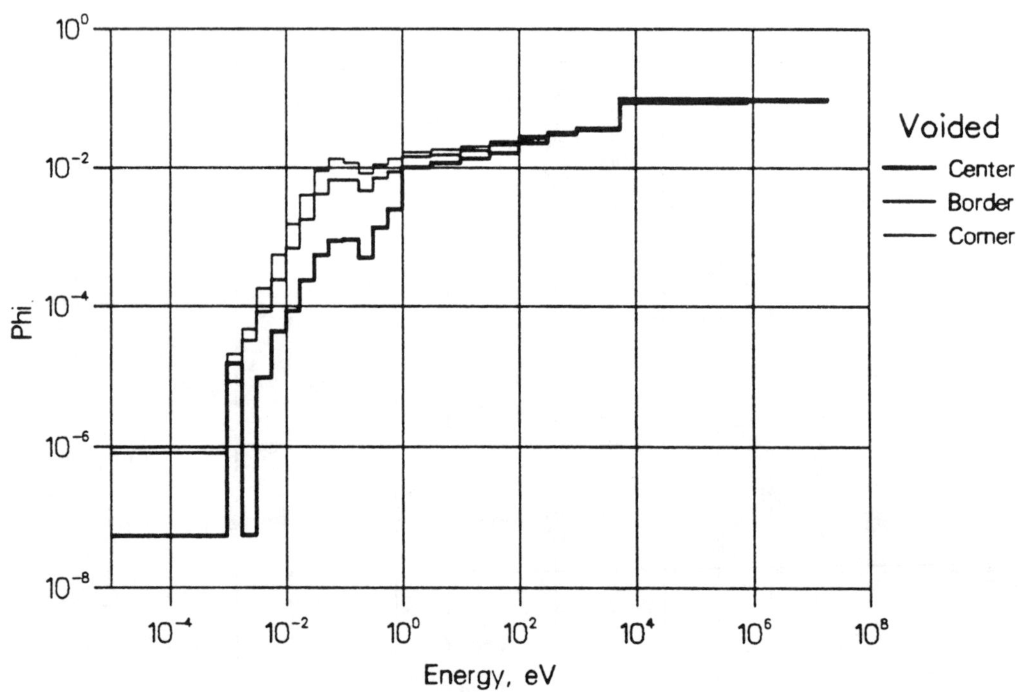

Fig. 15: Spectra in 3 cells

Uranium: CEA4

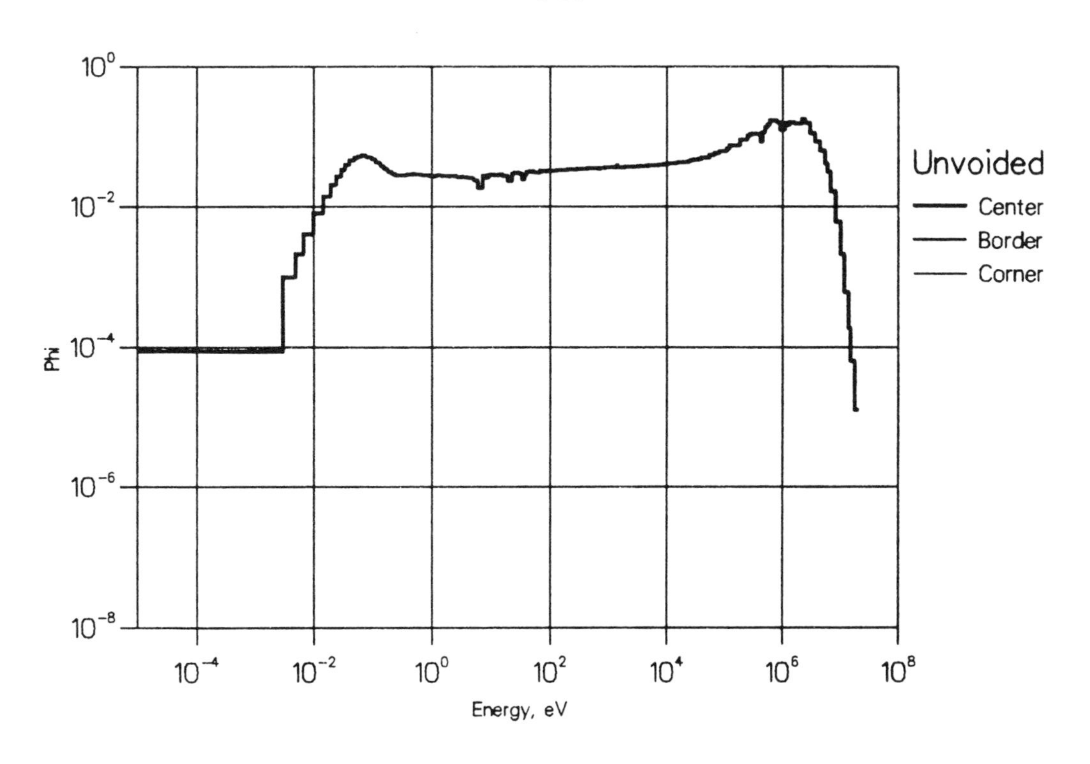

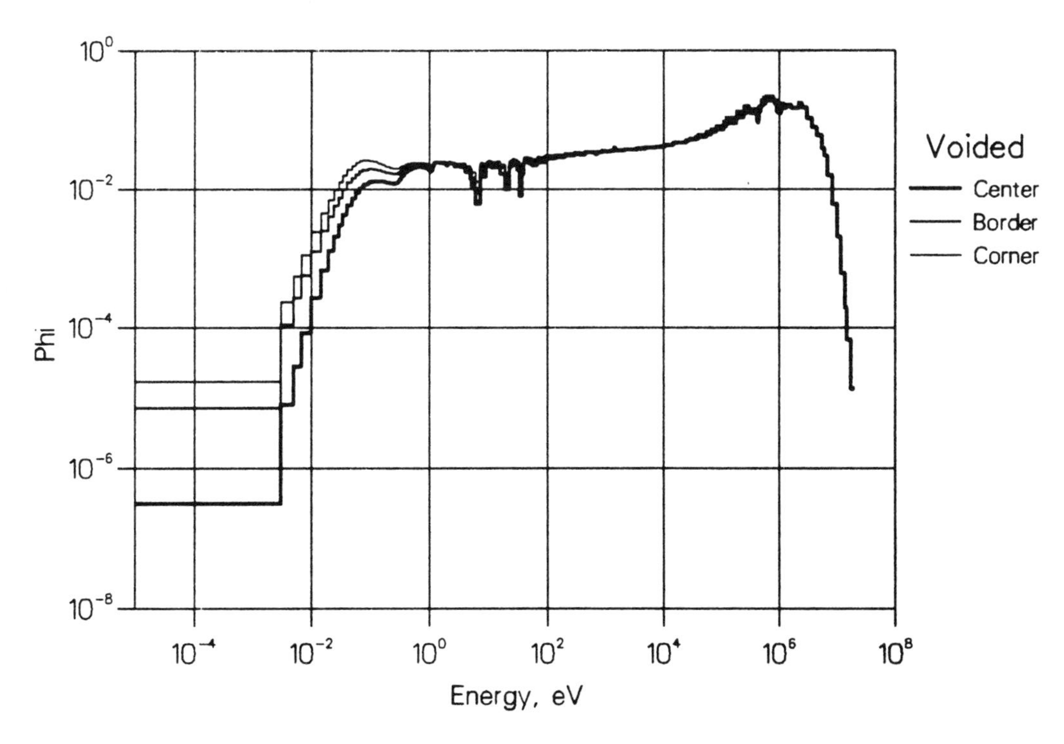

Fig. 16: Spectra in 3 cells

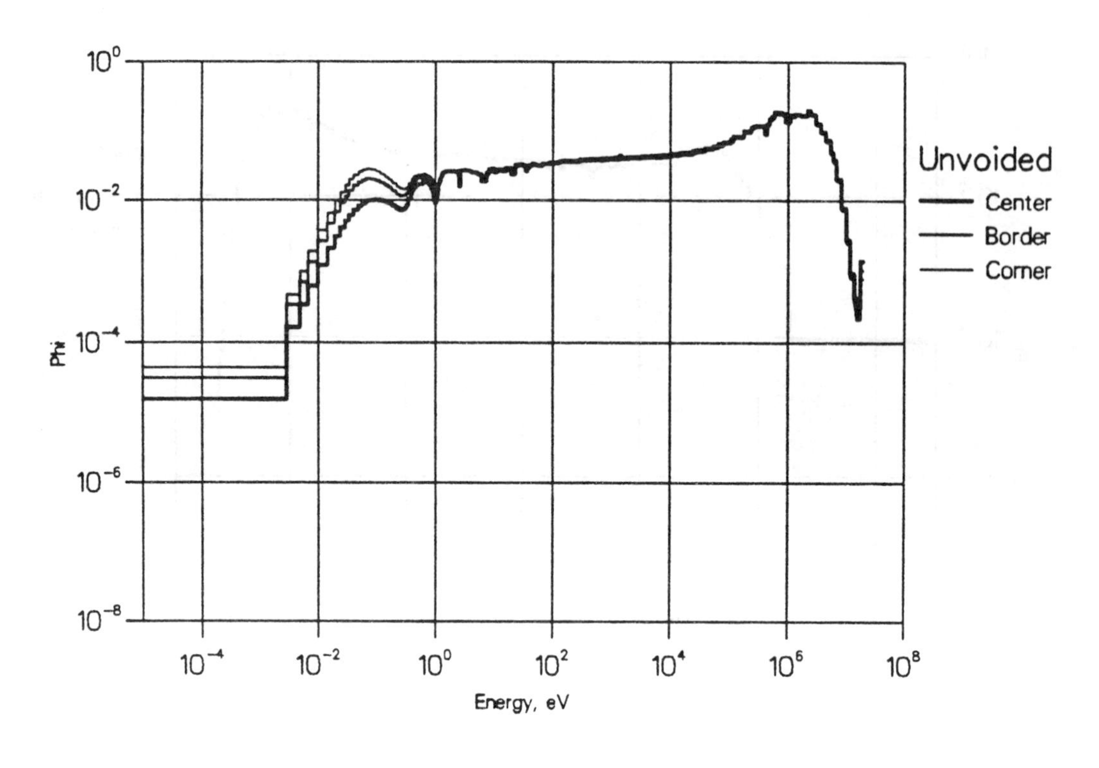

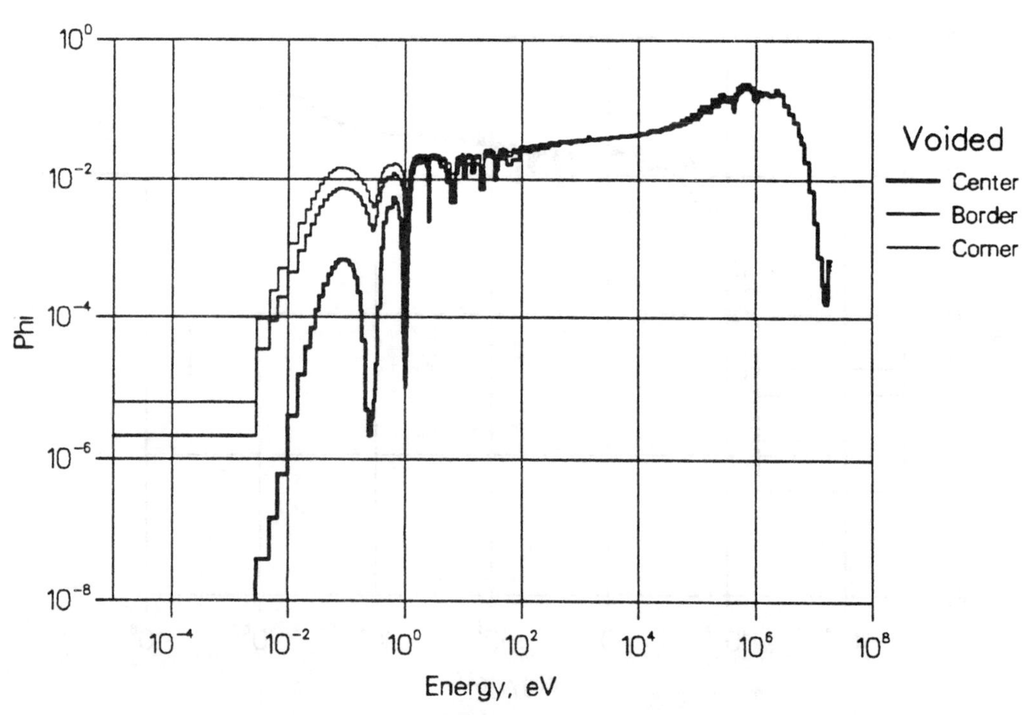

Fig. 17: Spectra in 3 cells

M–Mox: CEA4

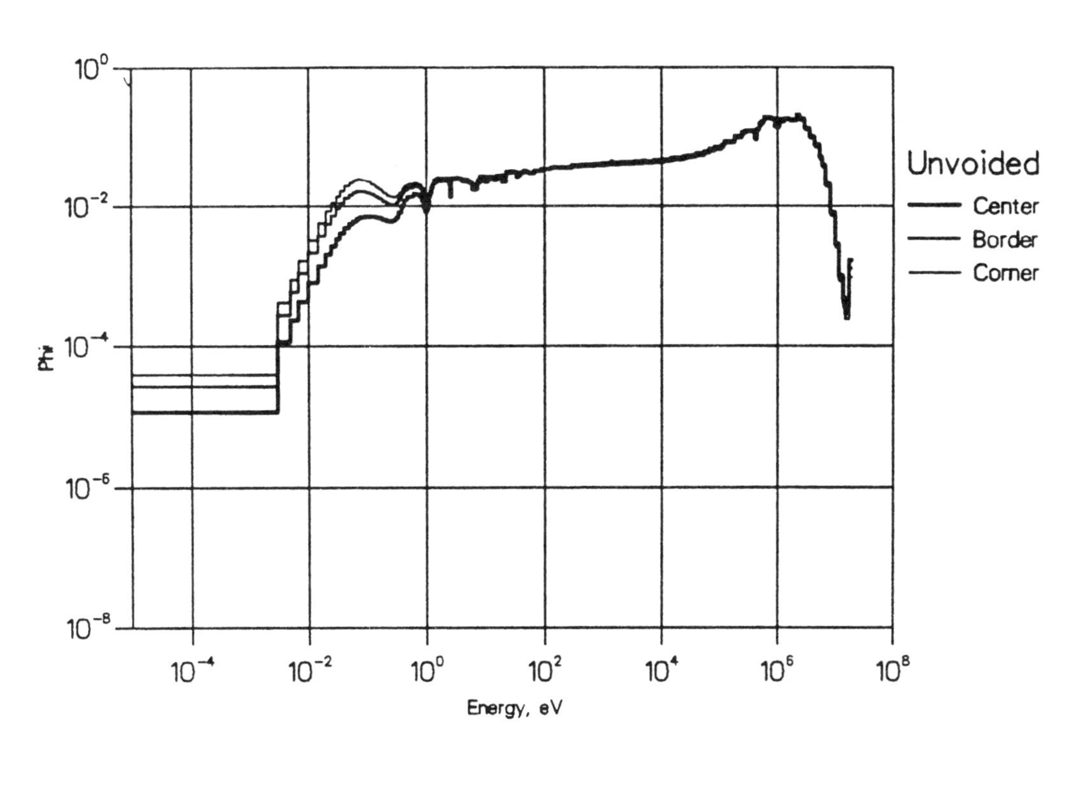

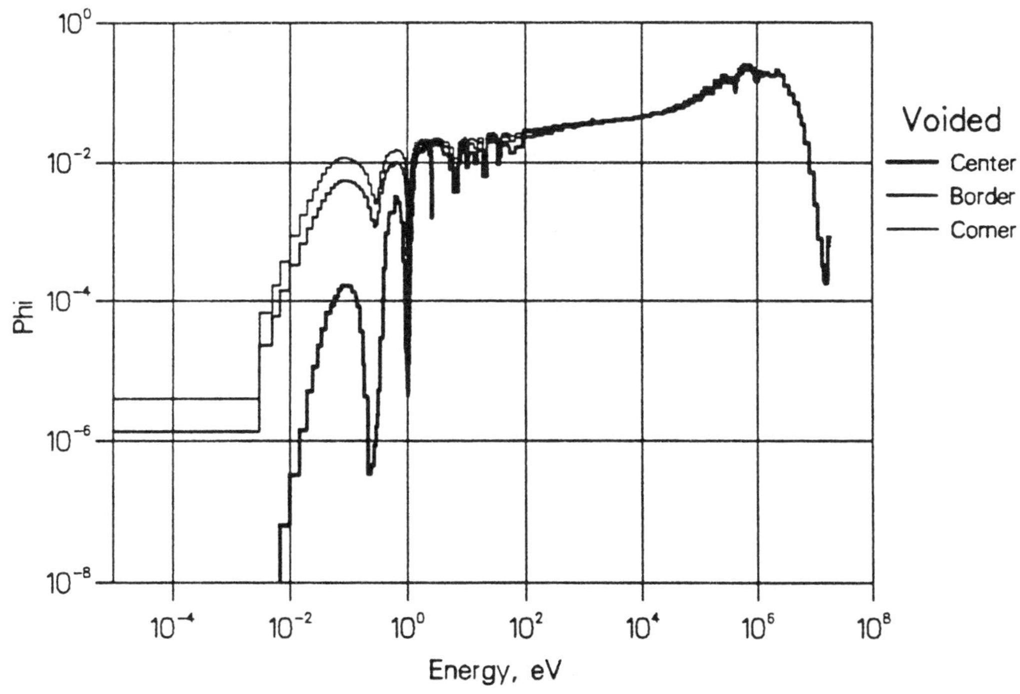

Fig. 18: Spectra in 3 cells

L–Mox: CEA4

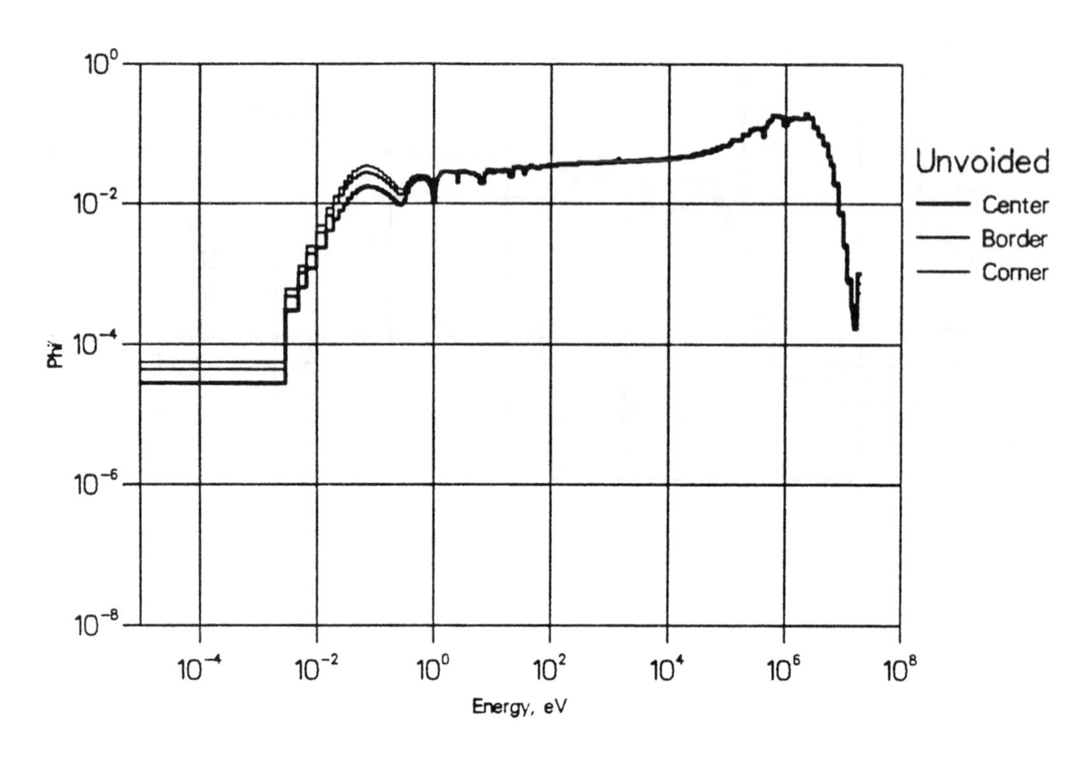

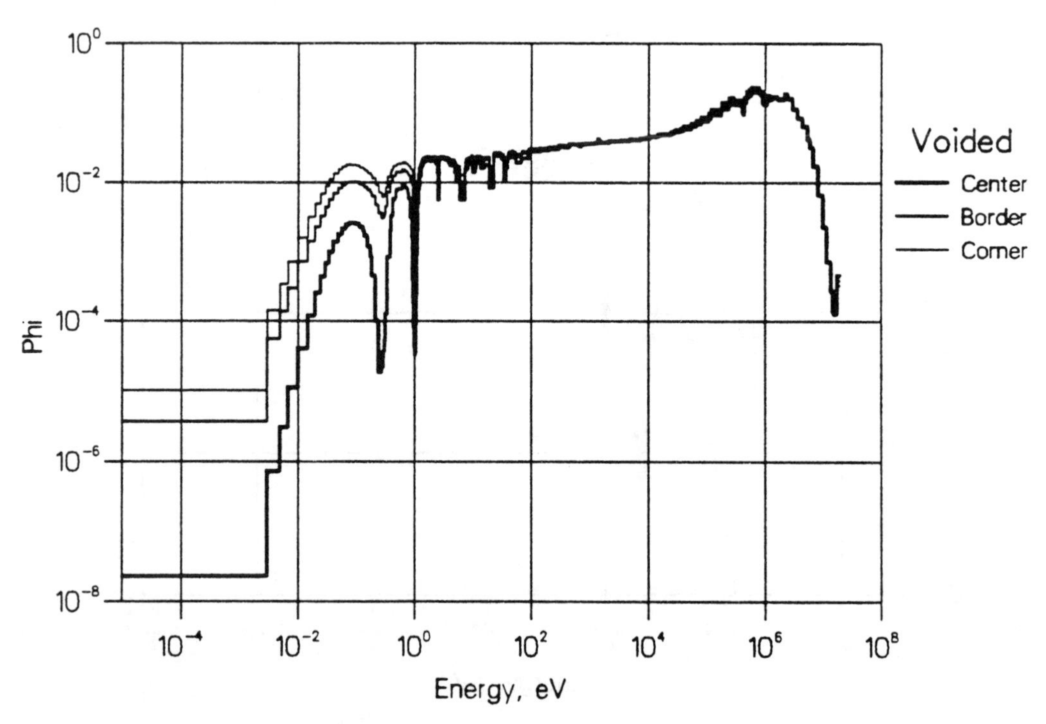

Fig. 19: Spectra in 3 cells

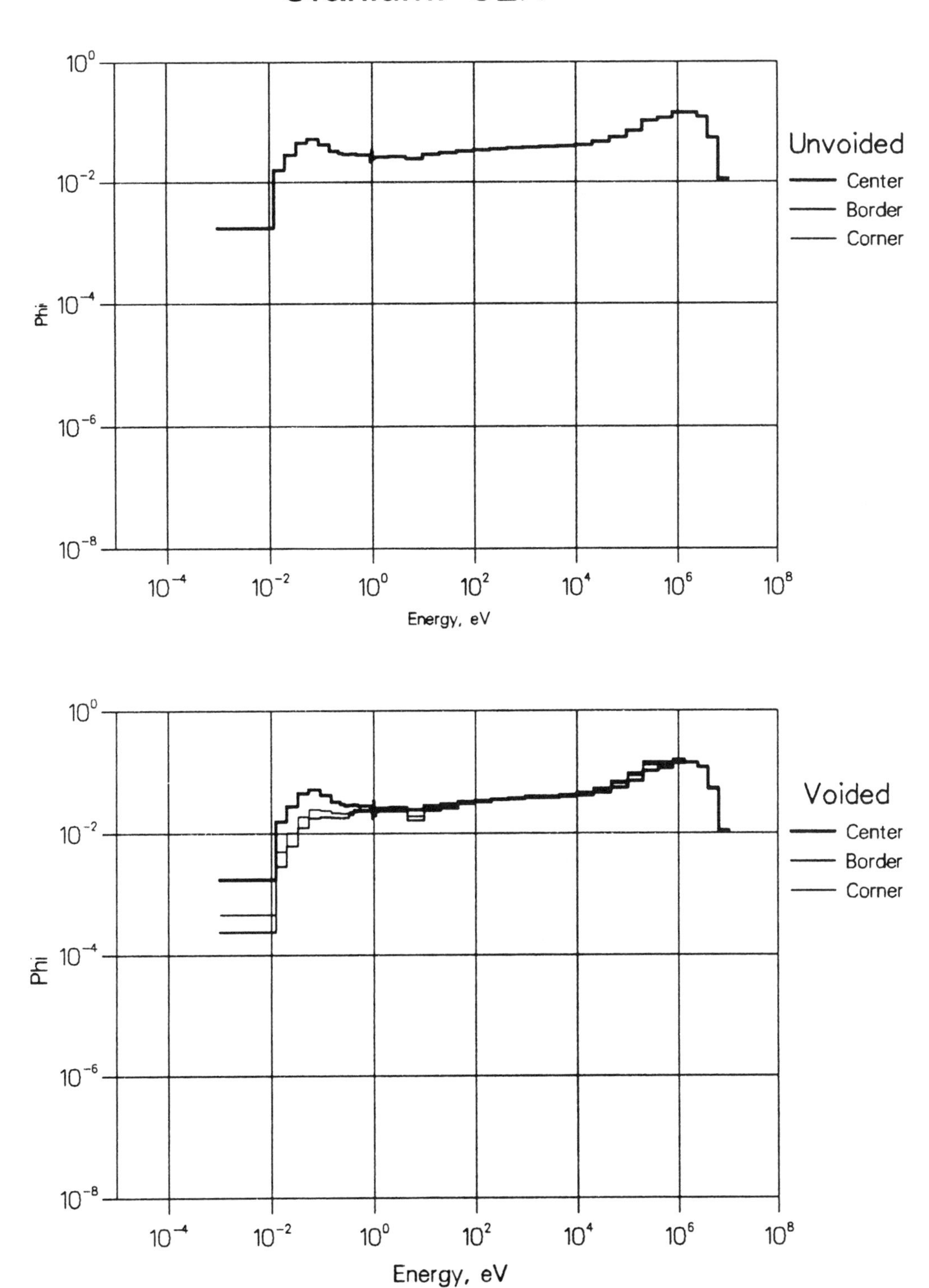

Fig. 20: Spectra in 3 cells

H–Mox: CEN

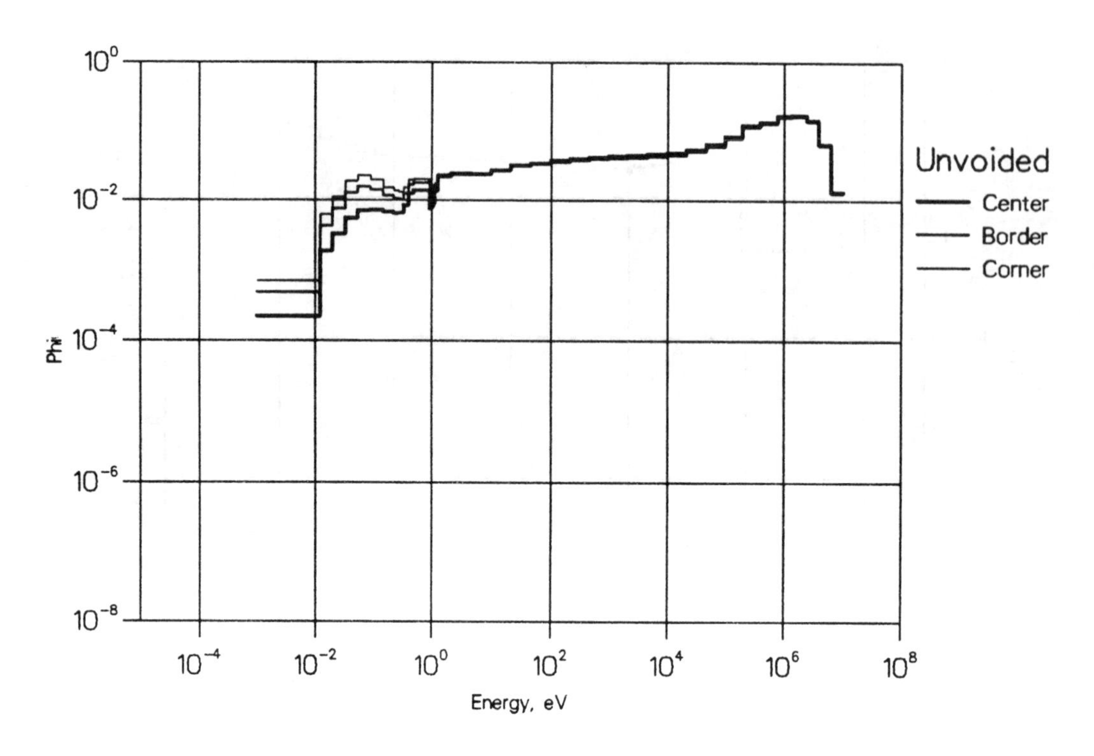

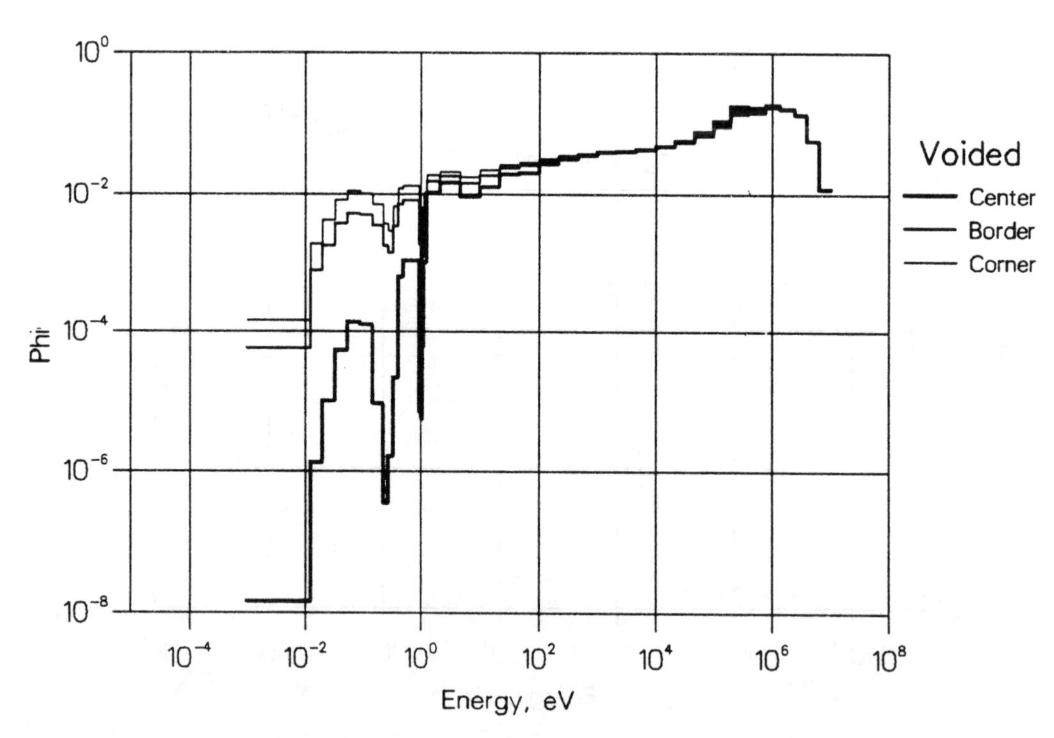

Fig. 21: Spectra in 3 cells

M–Mox: CEN

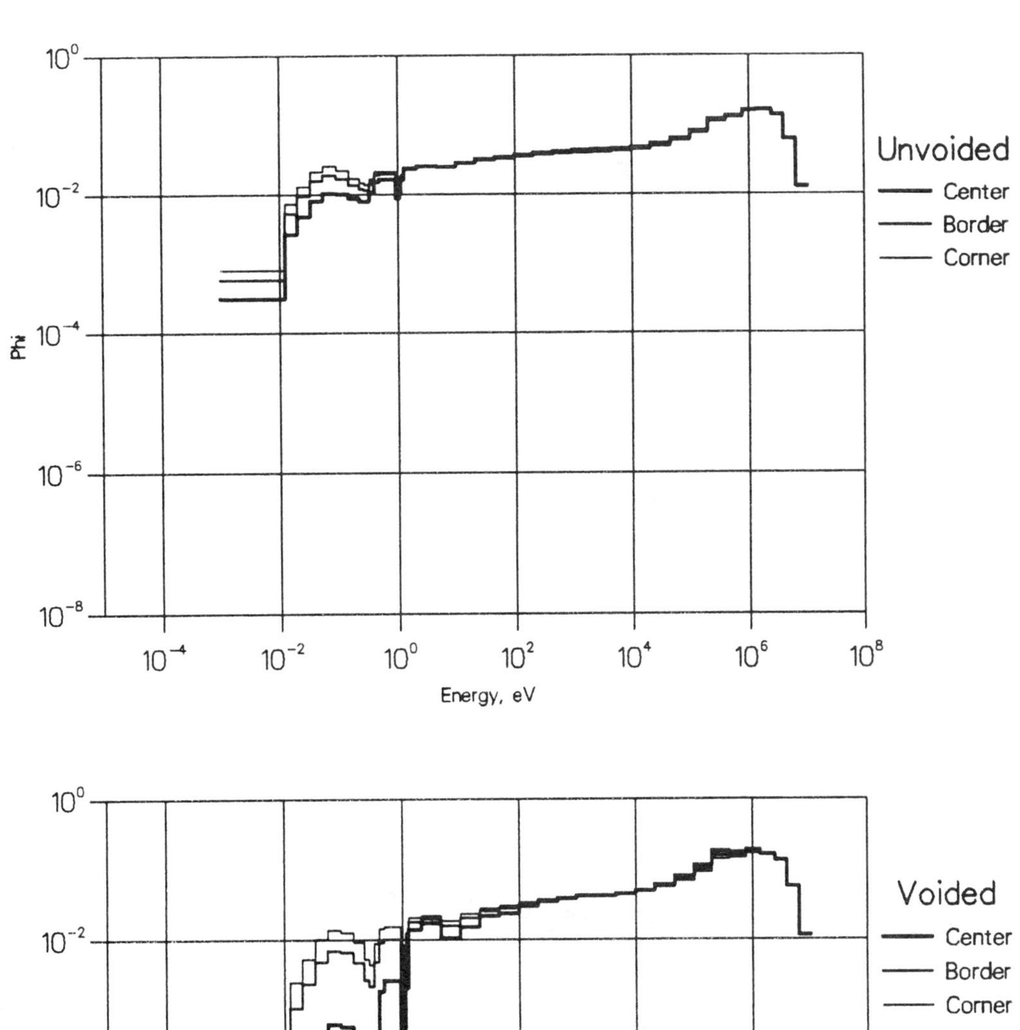

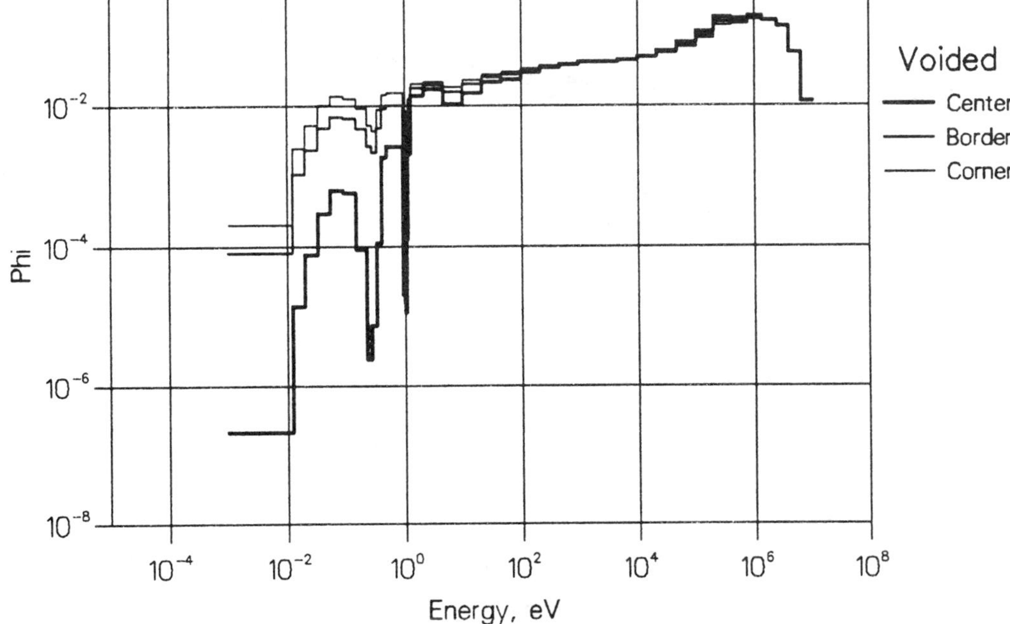

Fig. 22: Spectra in 3 cells

L–Mox: CEN

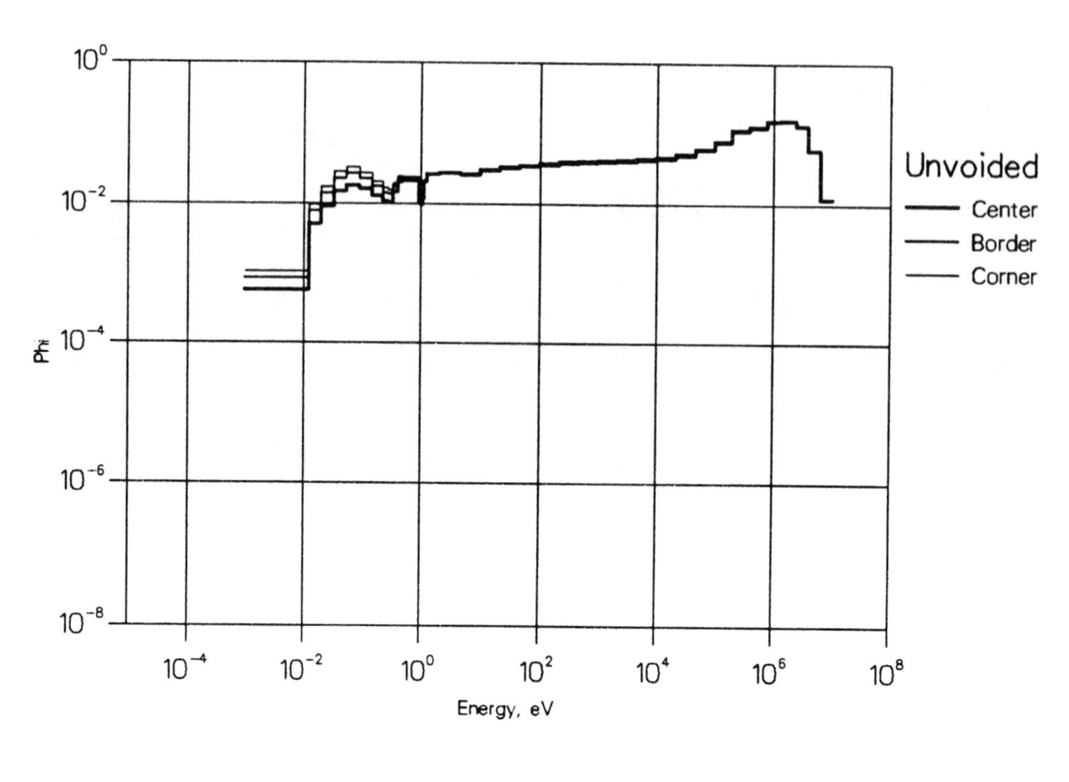

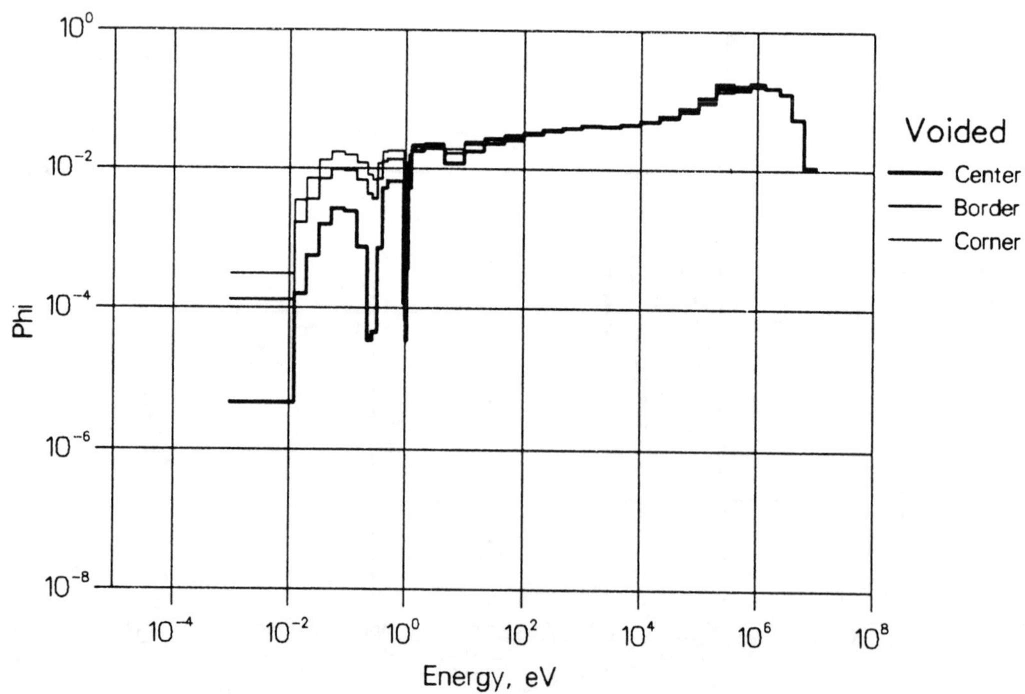

Fig. 23: Spectra in 3 cells

Uranium: IKE1

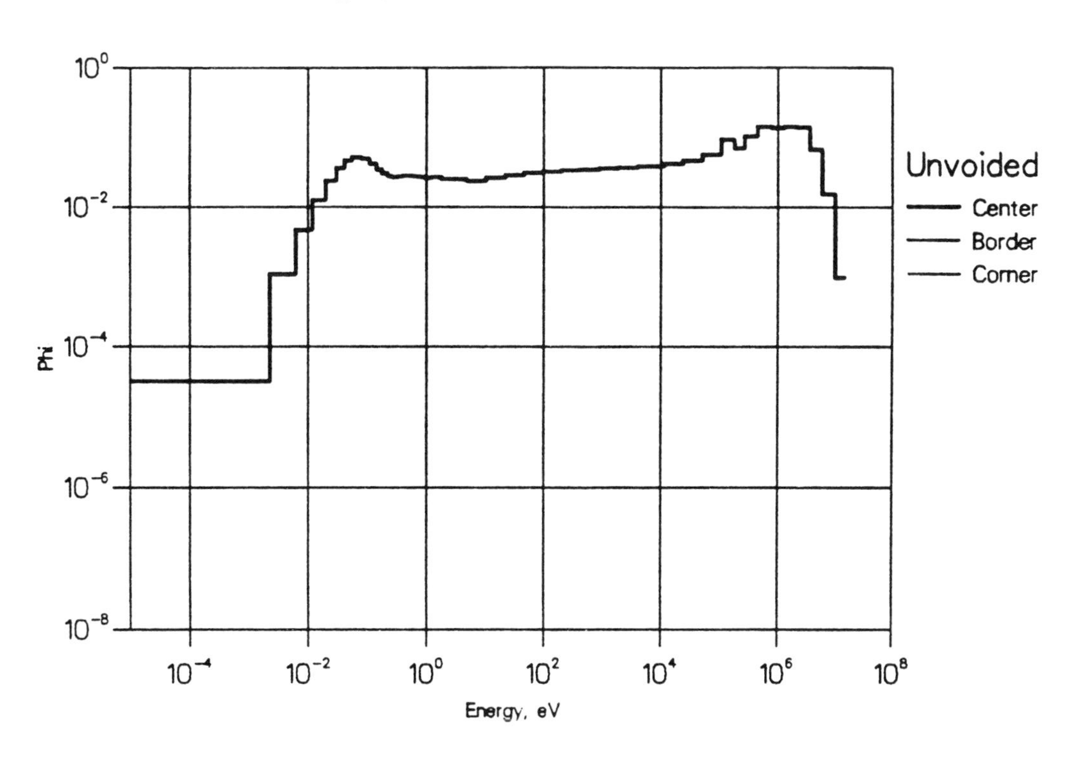

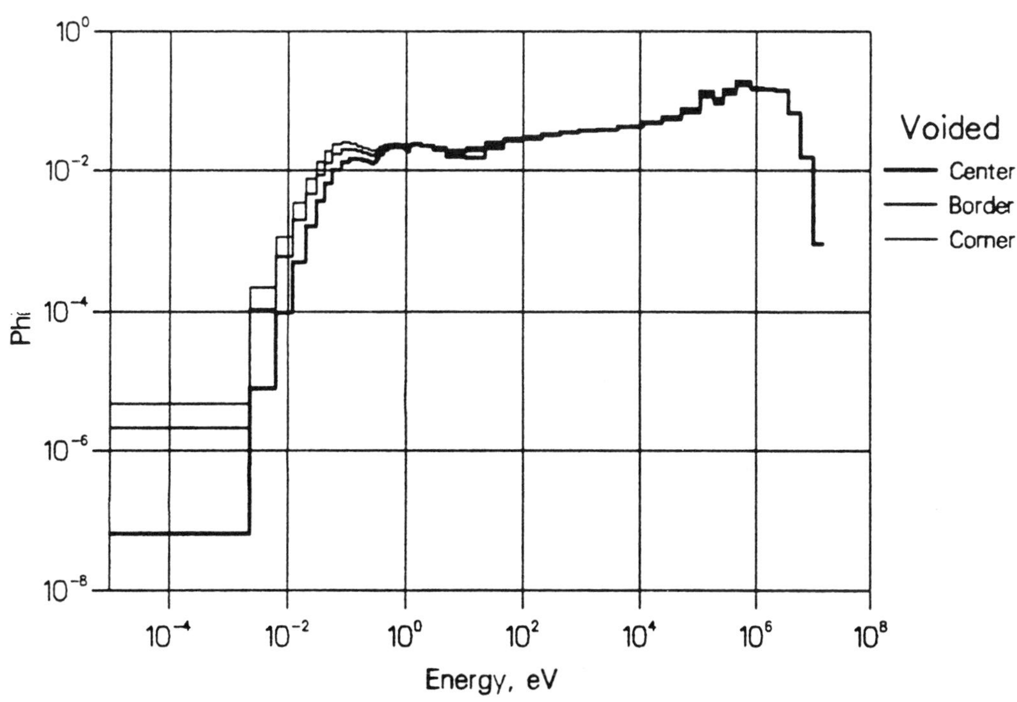

Fig. 24: Spectra in 3 cells

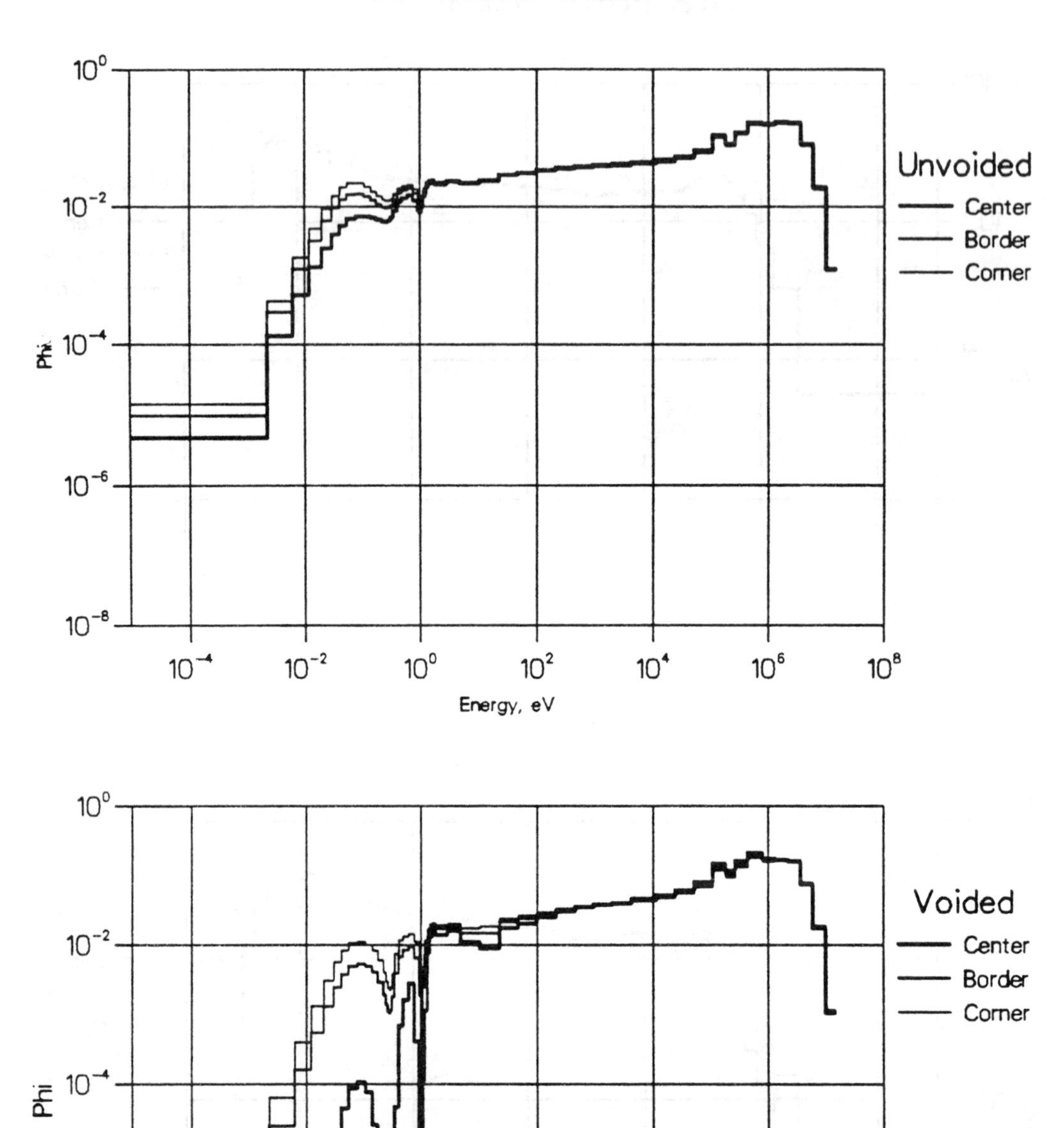

Fig. 25: Spectra in 3 cells.

M–Mox: IKE1

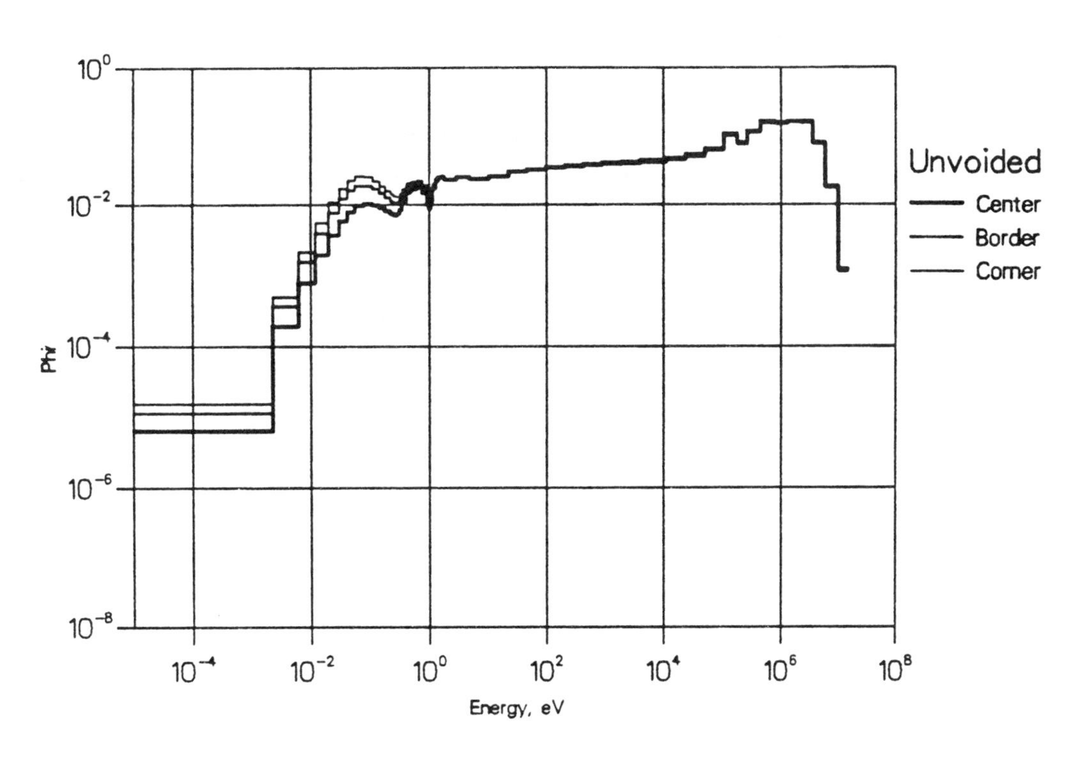

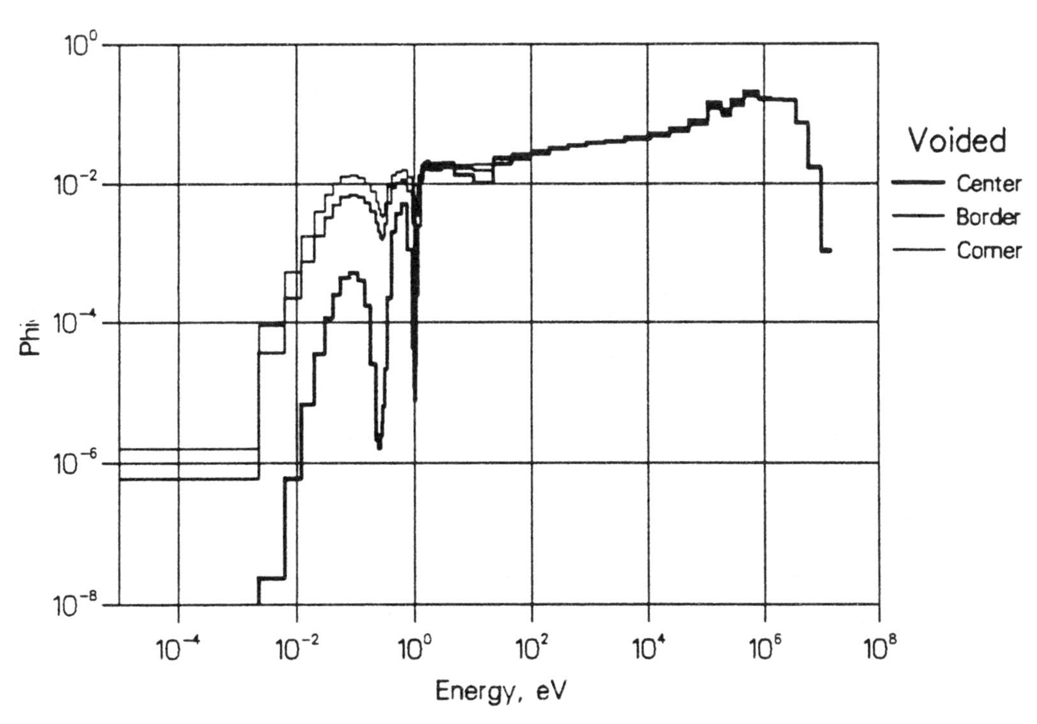

Fig. 26: Spectra in 3 cells

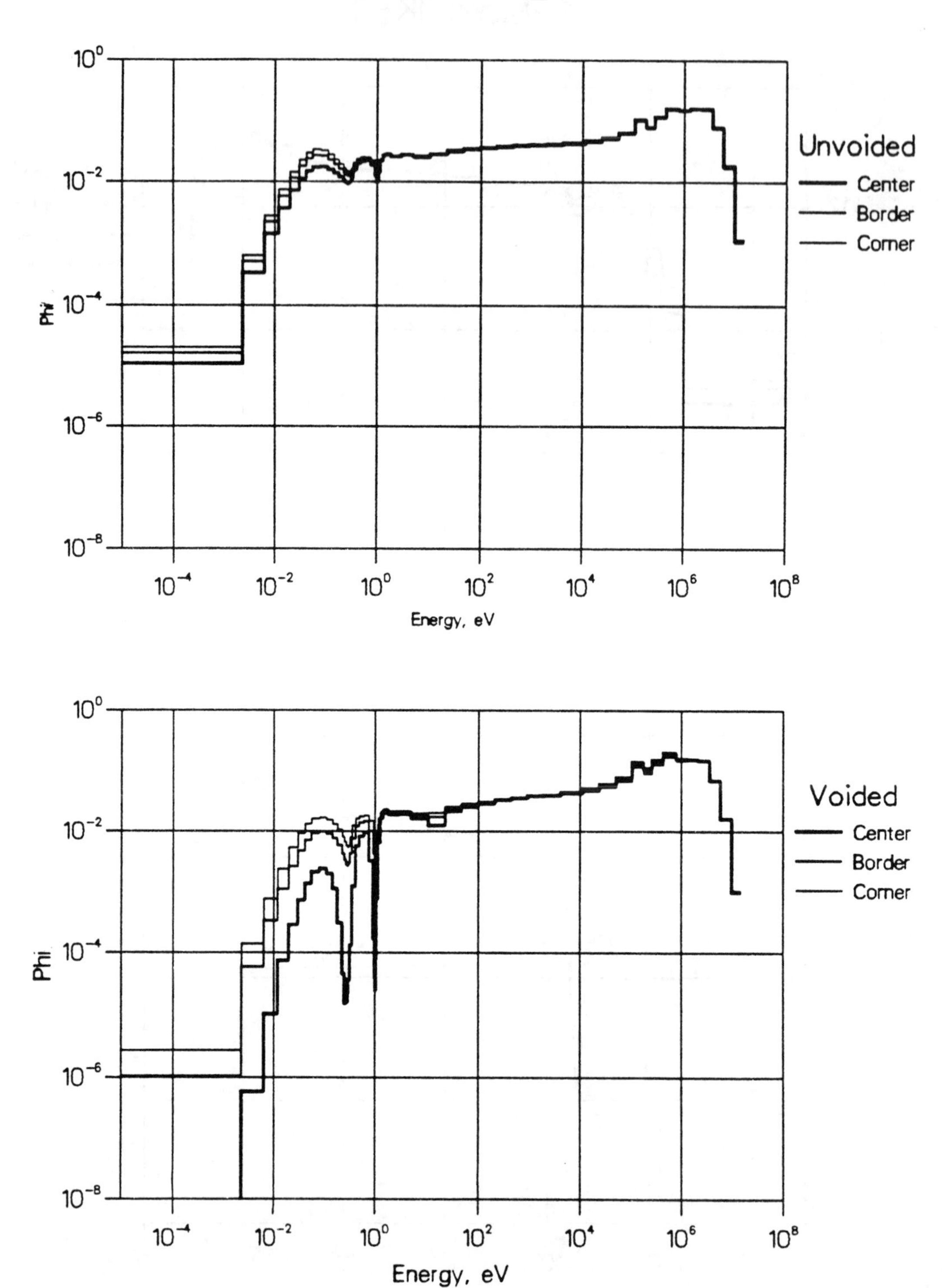

Fig. 27: Spectra in 3 cells

Uranium: IKE2

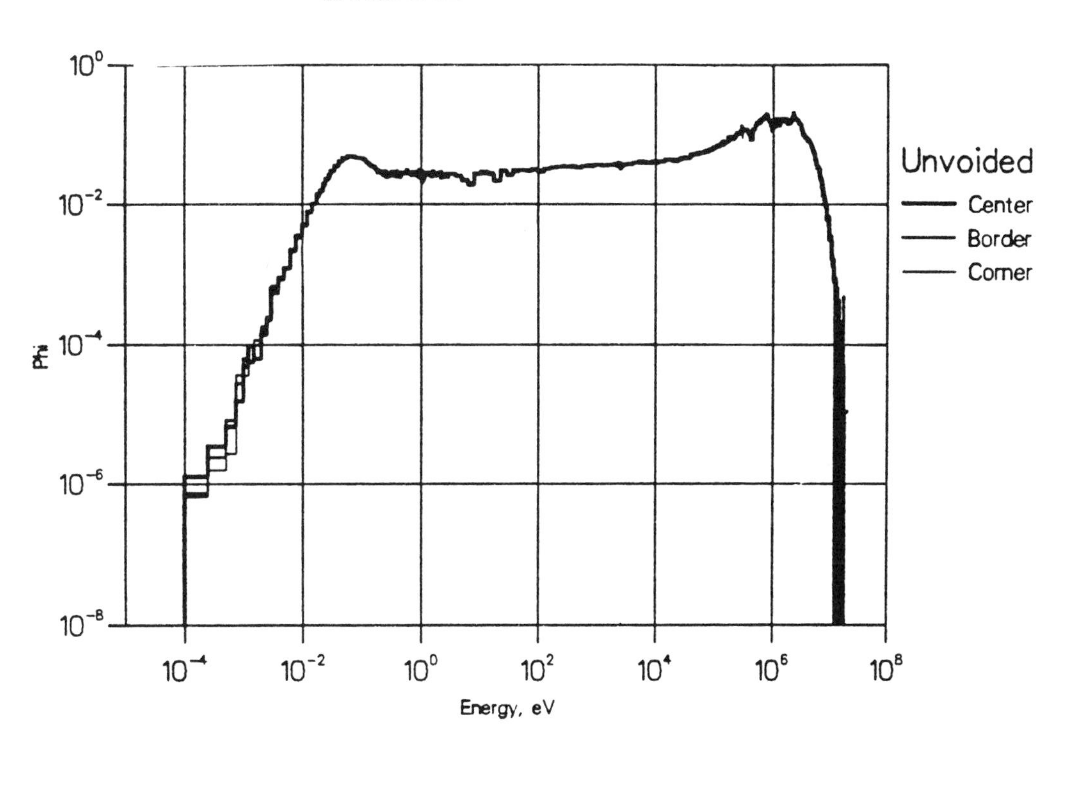

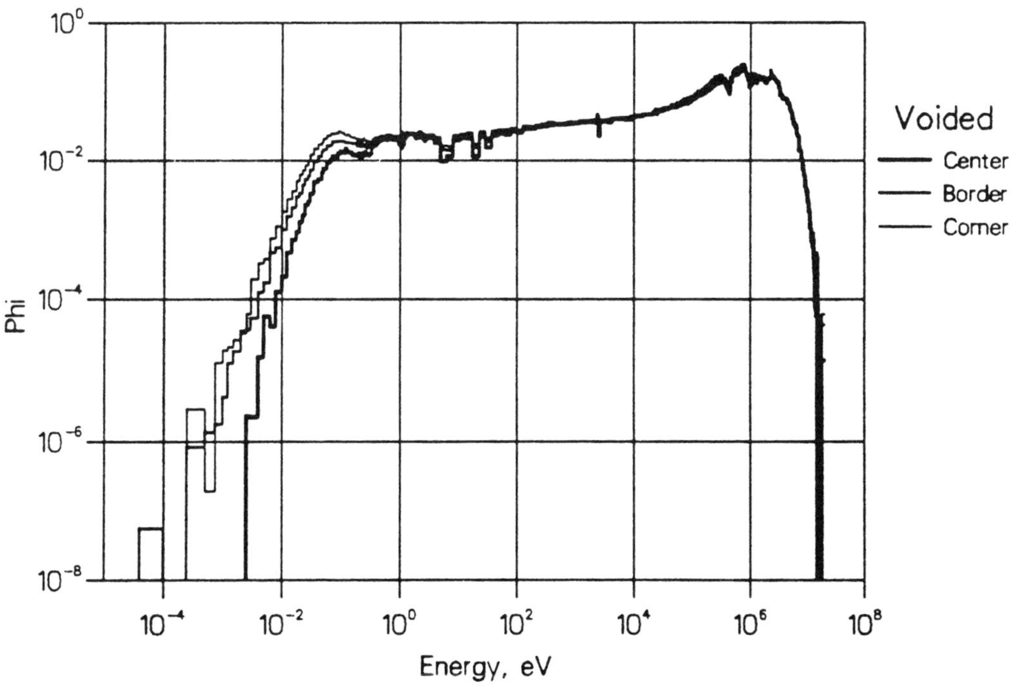

Fig. 28: Spectra in 3 cells

H–MOX: IKE2

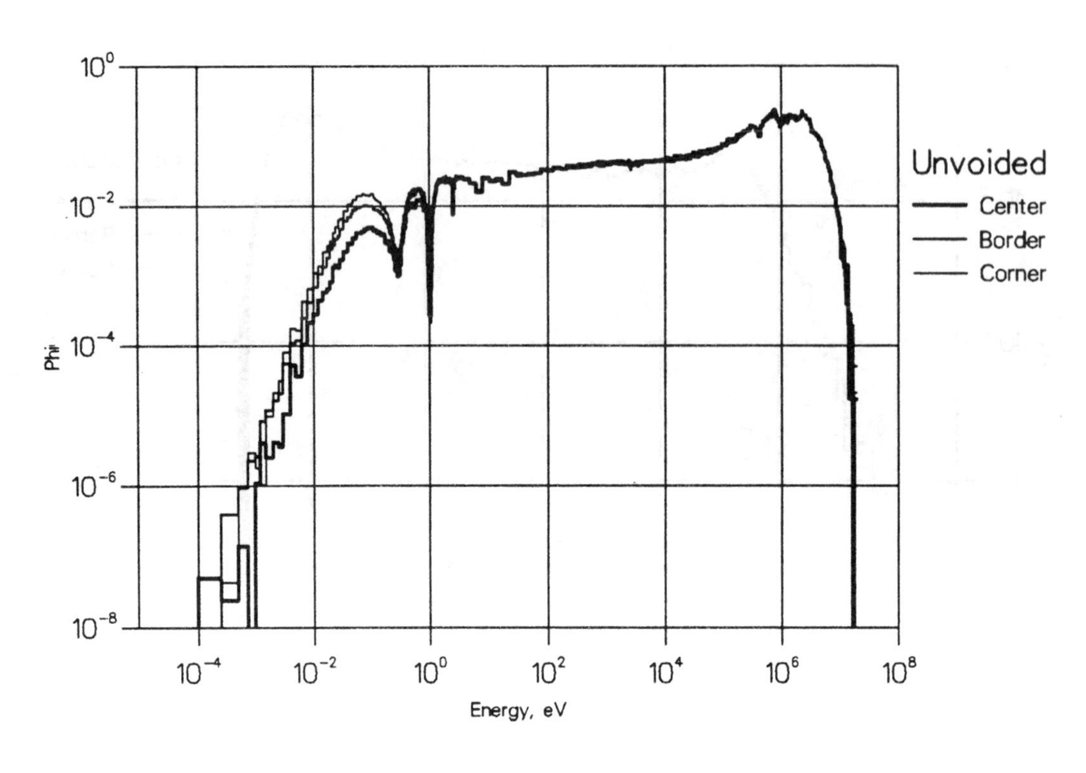

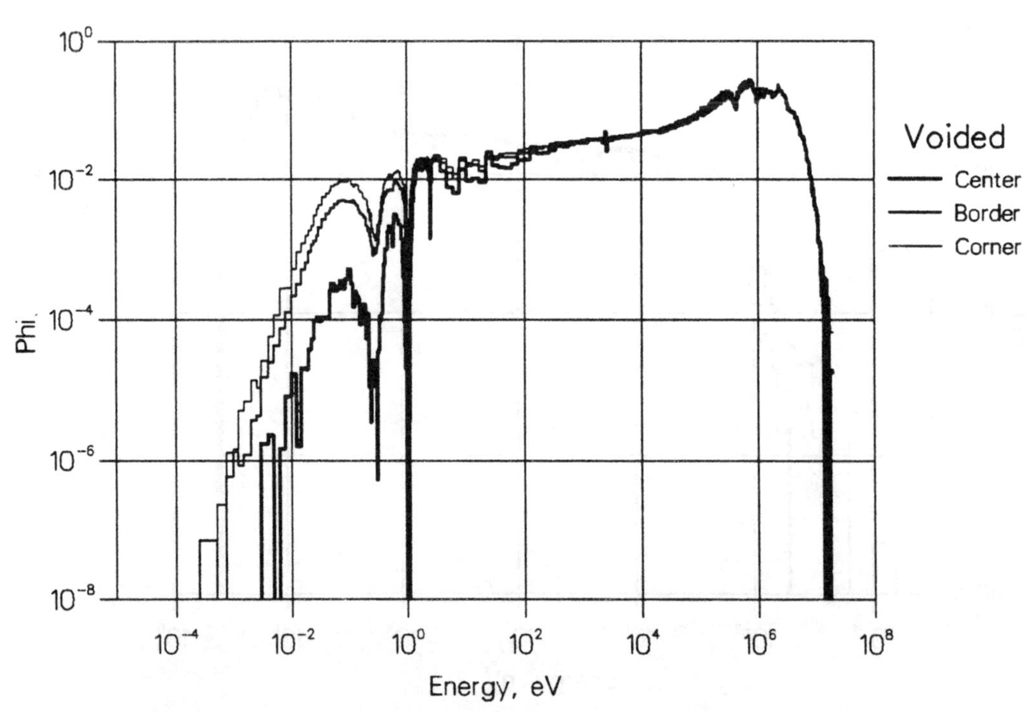

Fig. 29: Spectra in 3 cells.

M–Mox: IKE2

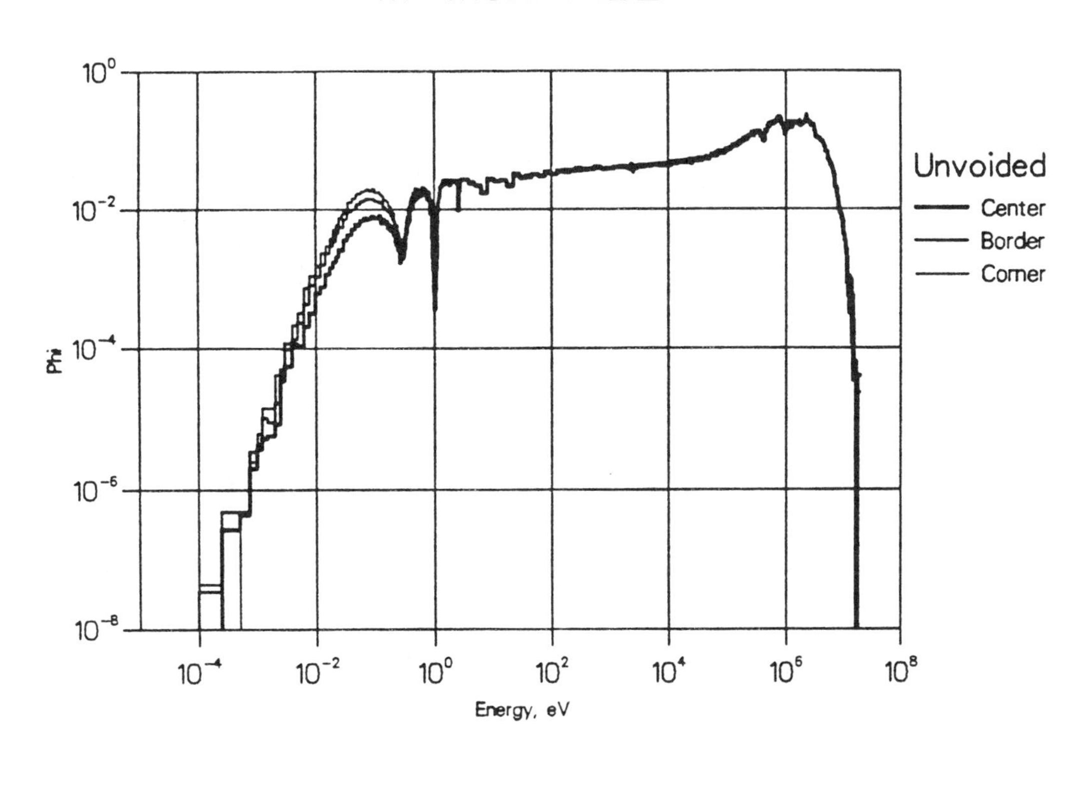

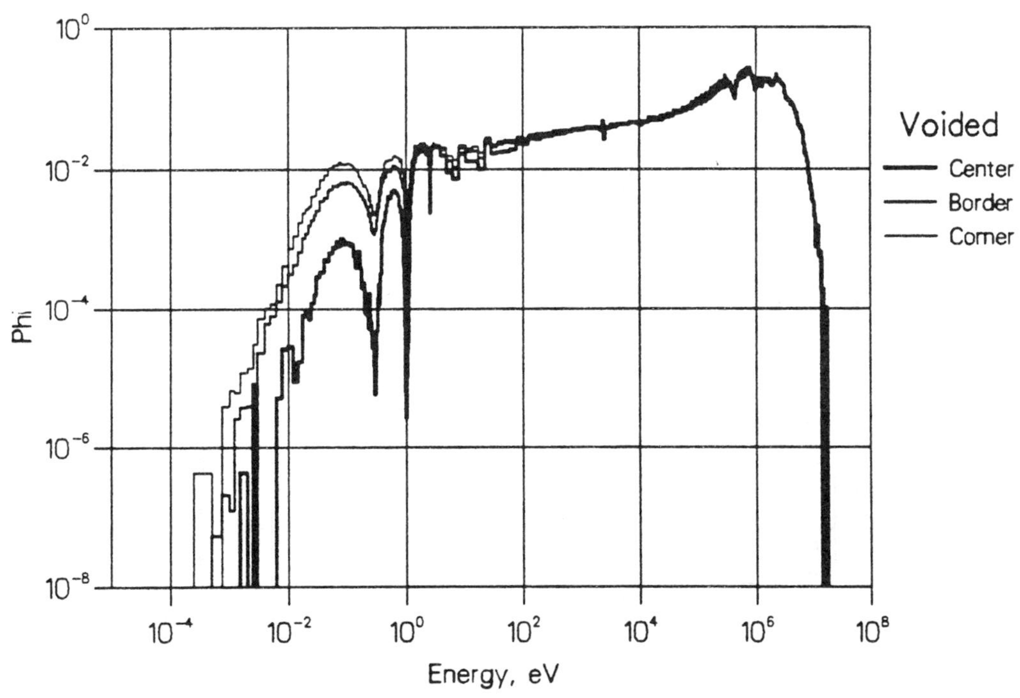

Fig. 30: Spectra in 3 cells,

L–Mox: IKE2

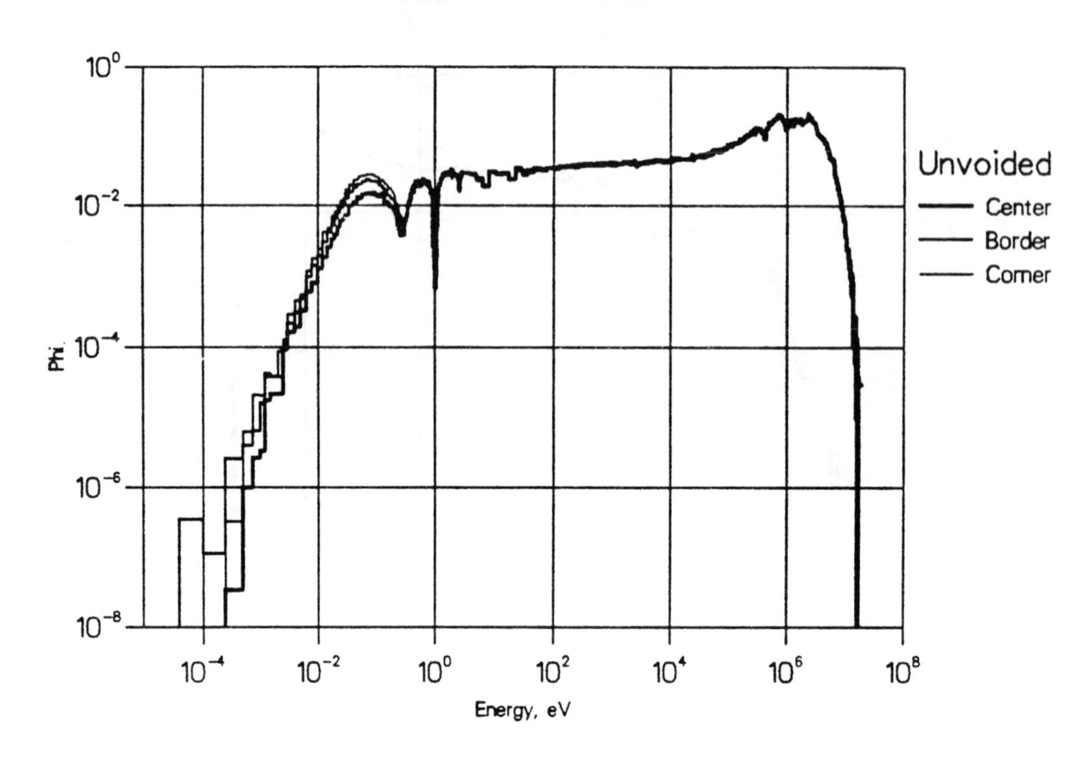

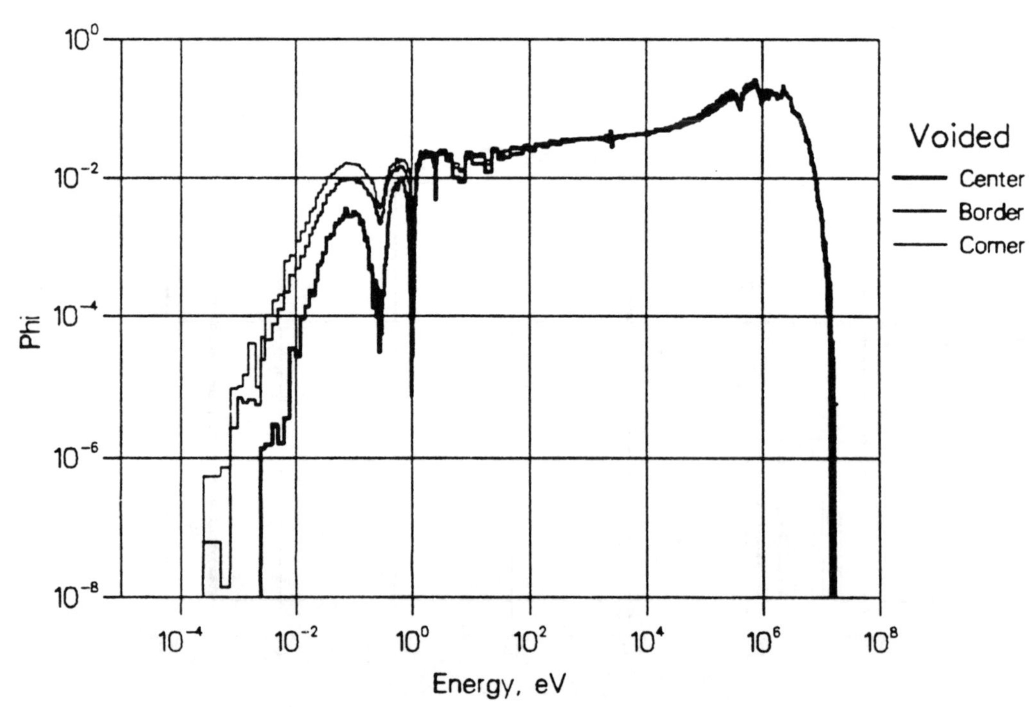

Fig. 31: Spectra in 3 cells

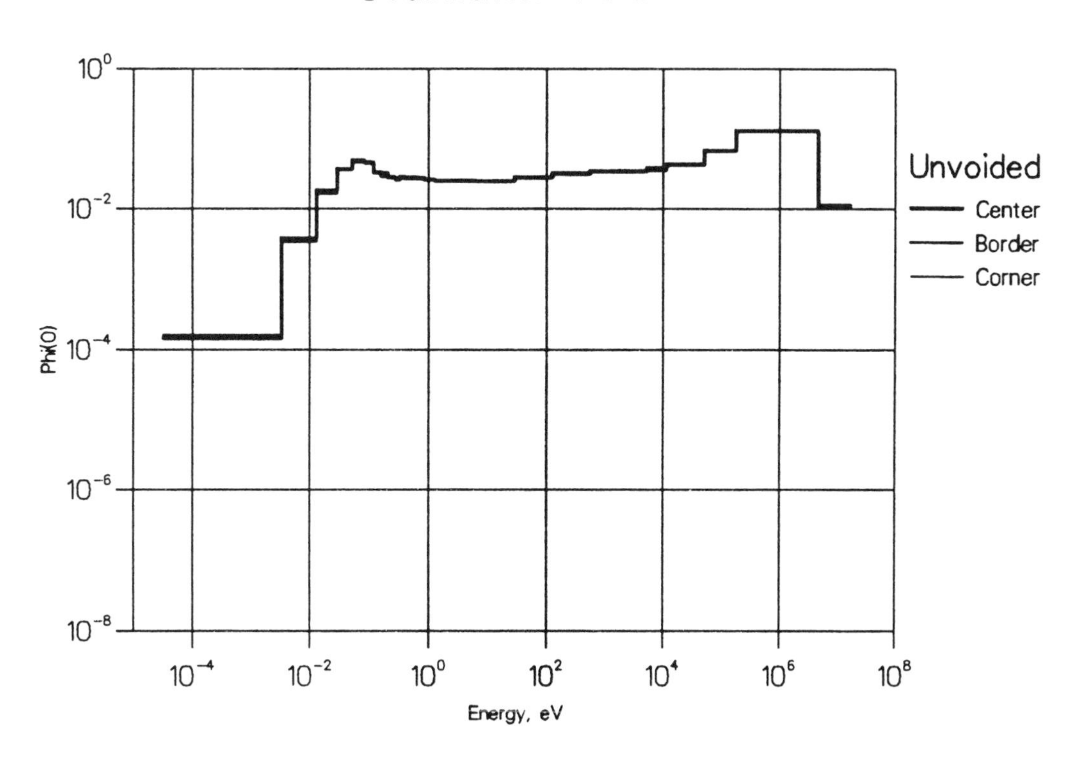

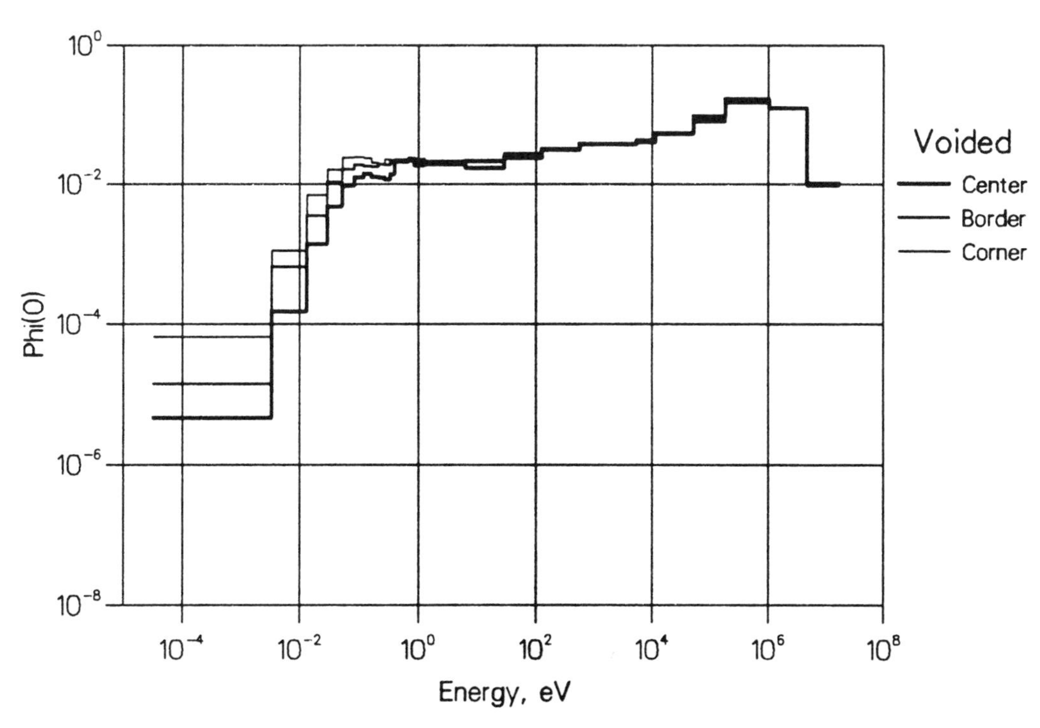

Fig. 32: Spectra in 3 cells

H–Mox: TOS

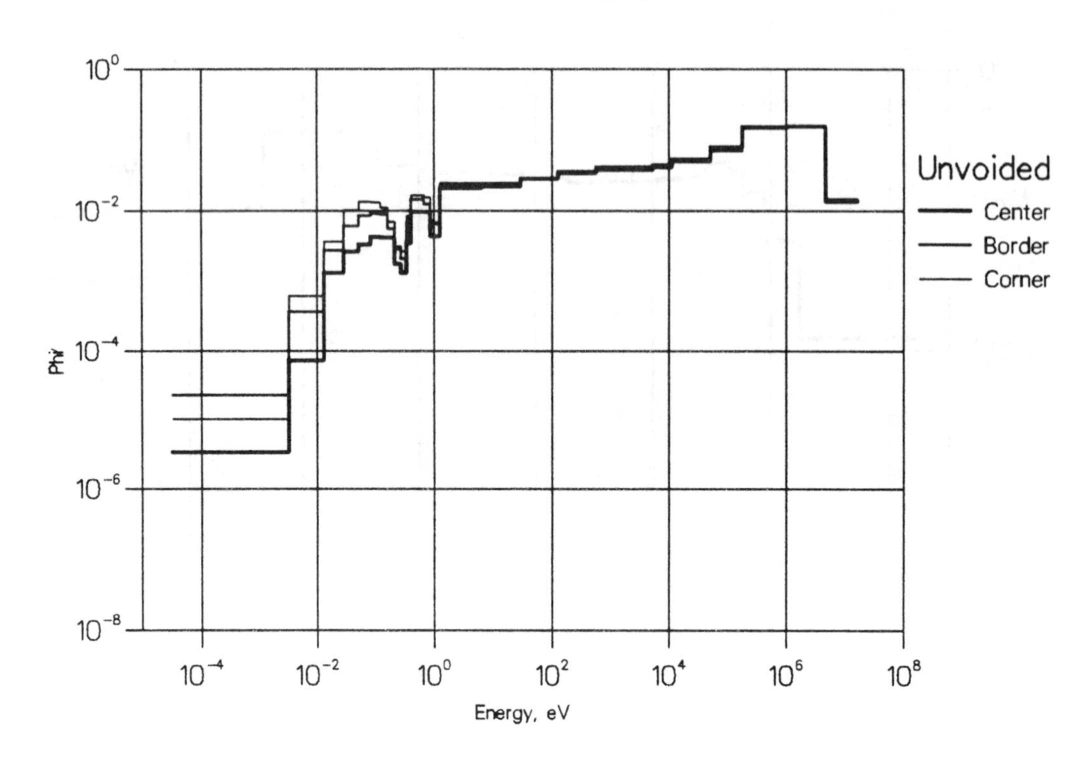

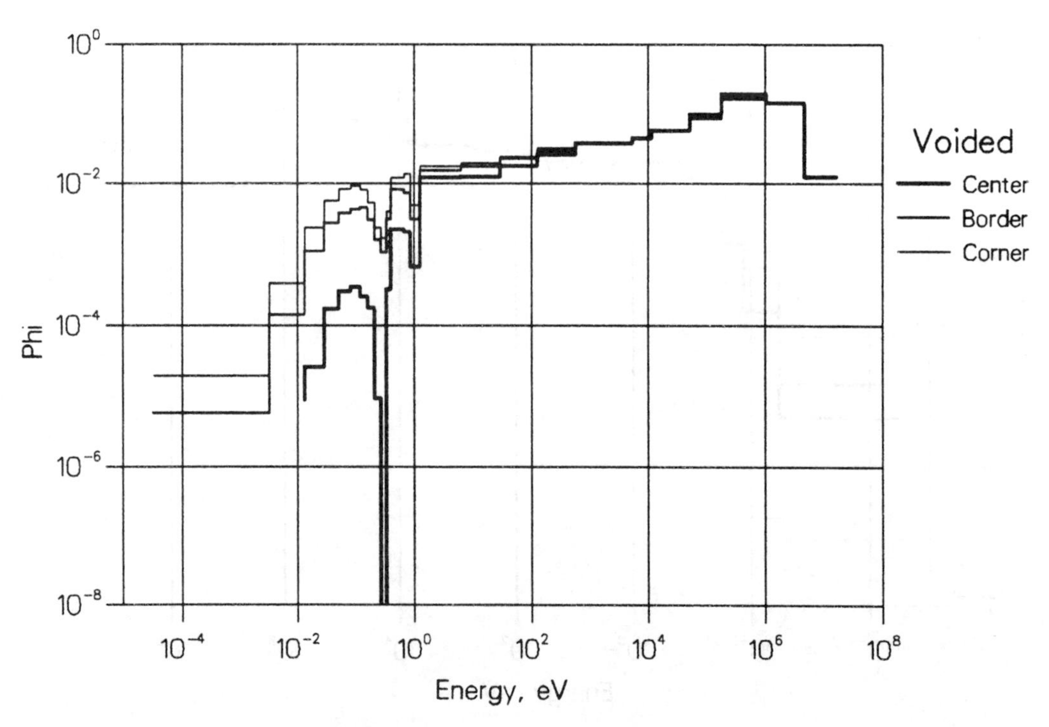

Fig. 33: Spectra in 3 cells

M–Mox: TOS

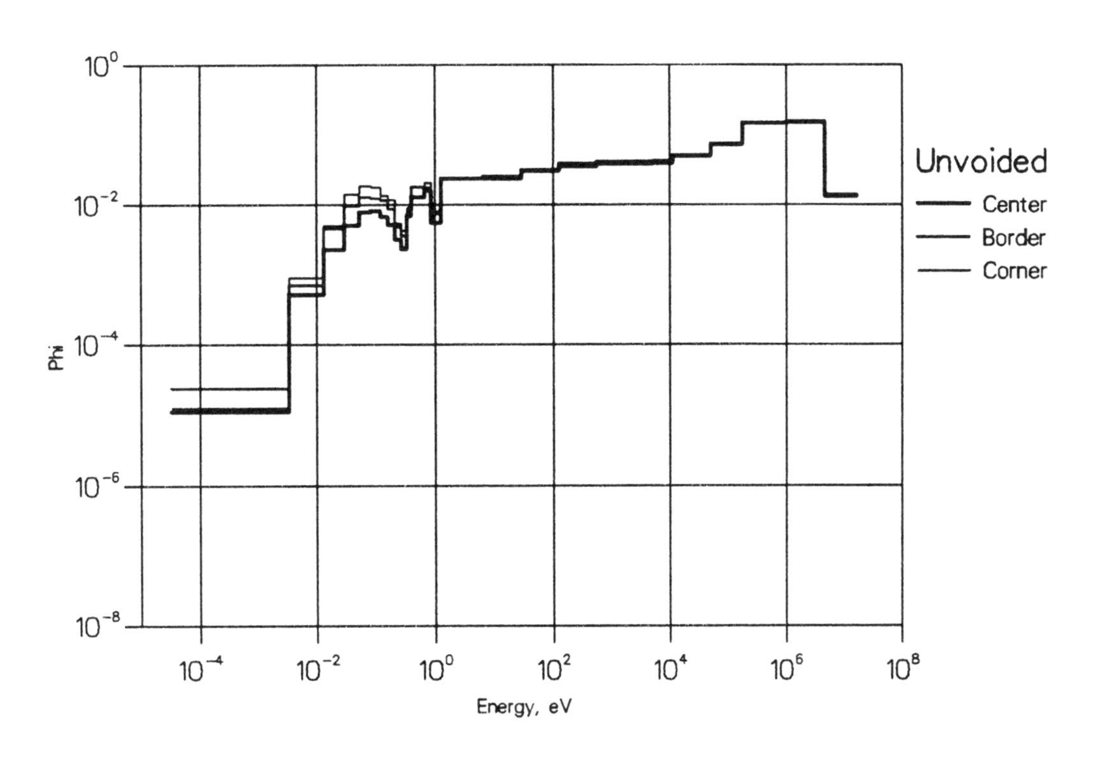

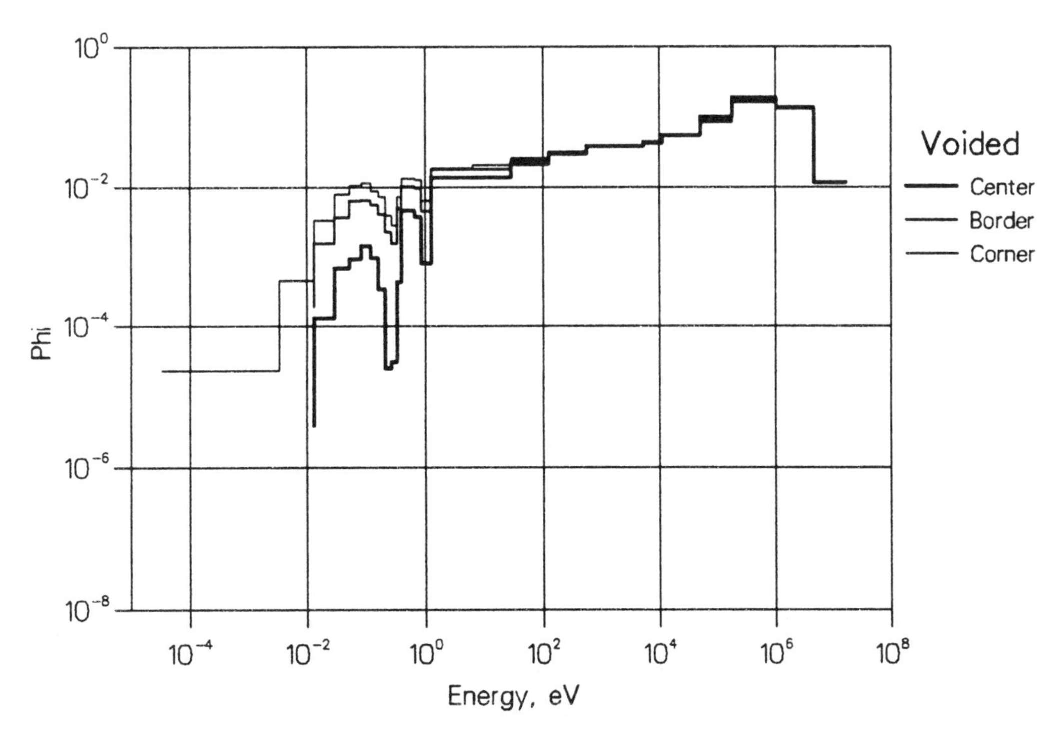

Fig. 34: Spectra in 3 cells

L–Mox: TOS

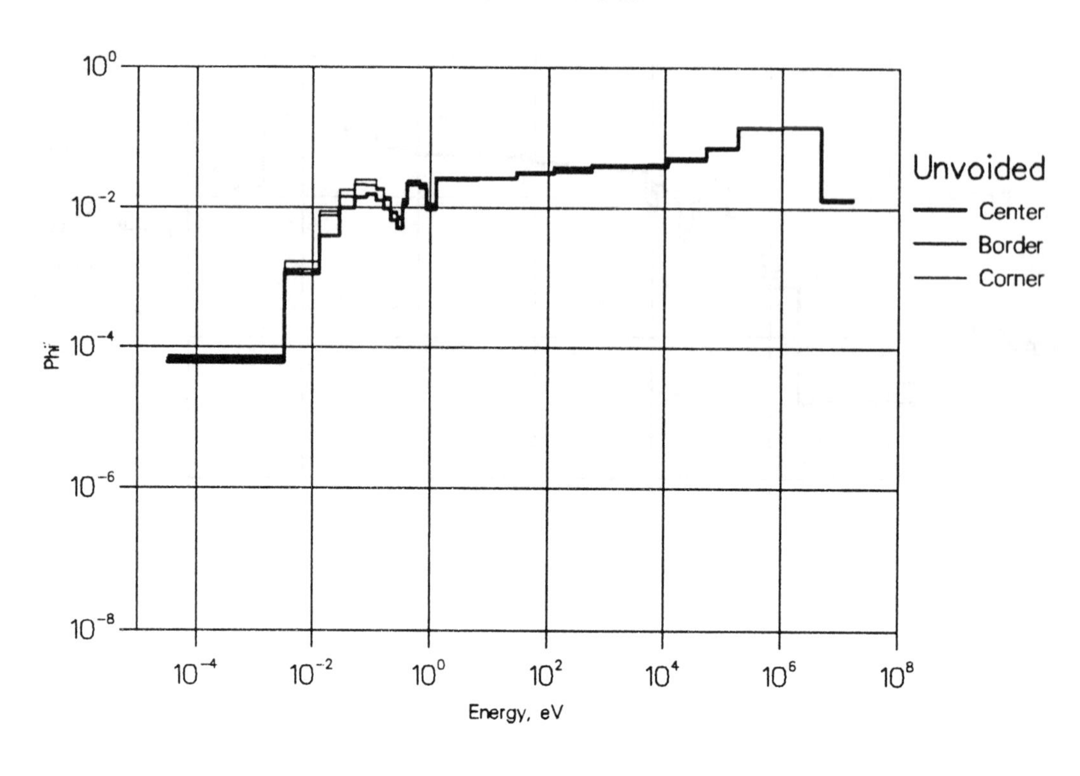

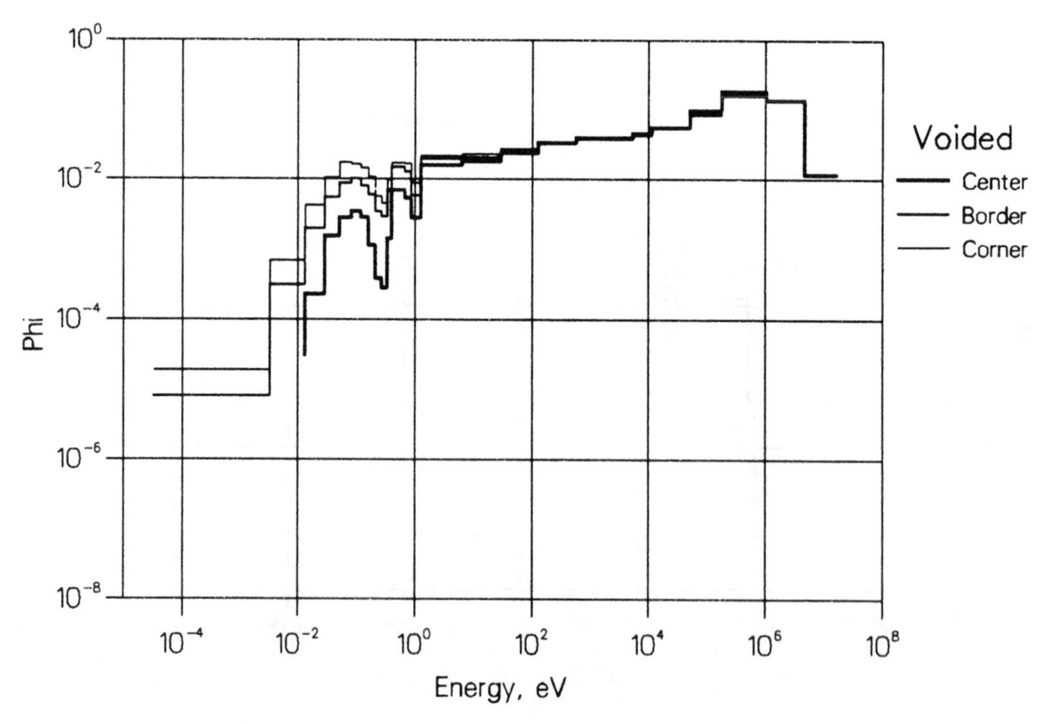

Fig. 35: Spectra in 3 cells

Appendix A

Benchmark specification

Void reactivity effect benchmark
A proposal for a computational exercise on void reactivity effect in MOX lattices

Th. Maldague (Belgonucléaire) and G. Minsart (SCK-CEN)

Introduction

It is known that the reactivity effect of voiding the central part of MOX-fuelled PWR lattices becomes less negative, and even positive, when the amount of Pu in the fuel is high. To check the ability of computing this void coefficient, a benchmark exercise is proposed. In a first stage, a simplified loading is considered, without leaking in order to remove the difficulties associated with this component of the neutron balance and to allow the use of only two-dimensional models. A possible further step could be the analysis of a real configuration.

It is proposed that participants use their own procedure of computation (computer programs and cross-section libraries).

Geometrical data

A "macrocell" made of a 30×30 unit cells array is considered with reflective boundary conditions on all faces and an axial buckling = 0.0; the lattice pitch is 1.26 cm (see Figure A-1). It is divided into two zones:

- An outer zone containing UO_2 fuel rods,
- An inner zone in which several fuel compositions are to be introduced, with and without moderator.

The geometry of the unit cells is:

- Square lattice pitch: 1.26 cm,
- Fuel pellet outer radius: 0.4095 cm,
- Cladding outer radius: 0.4750 cm.
  The gap is smeared with the cladding.
- Fuel and water temperature: 20°C.

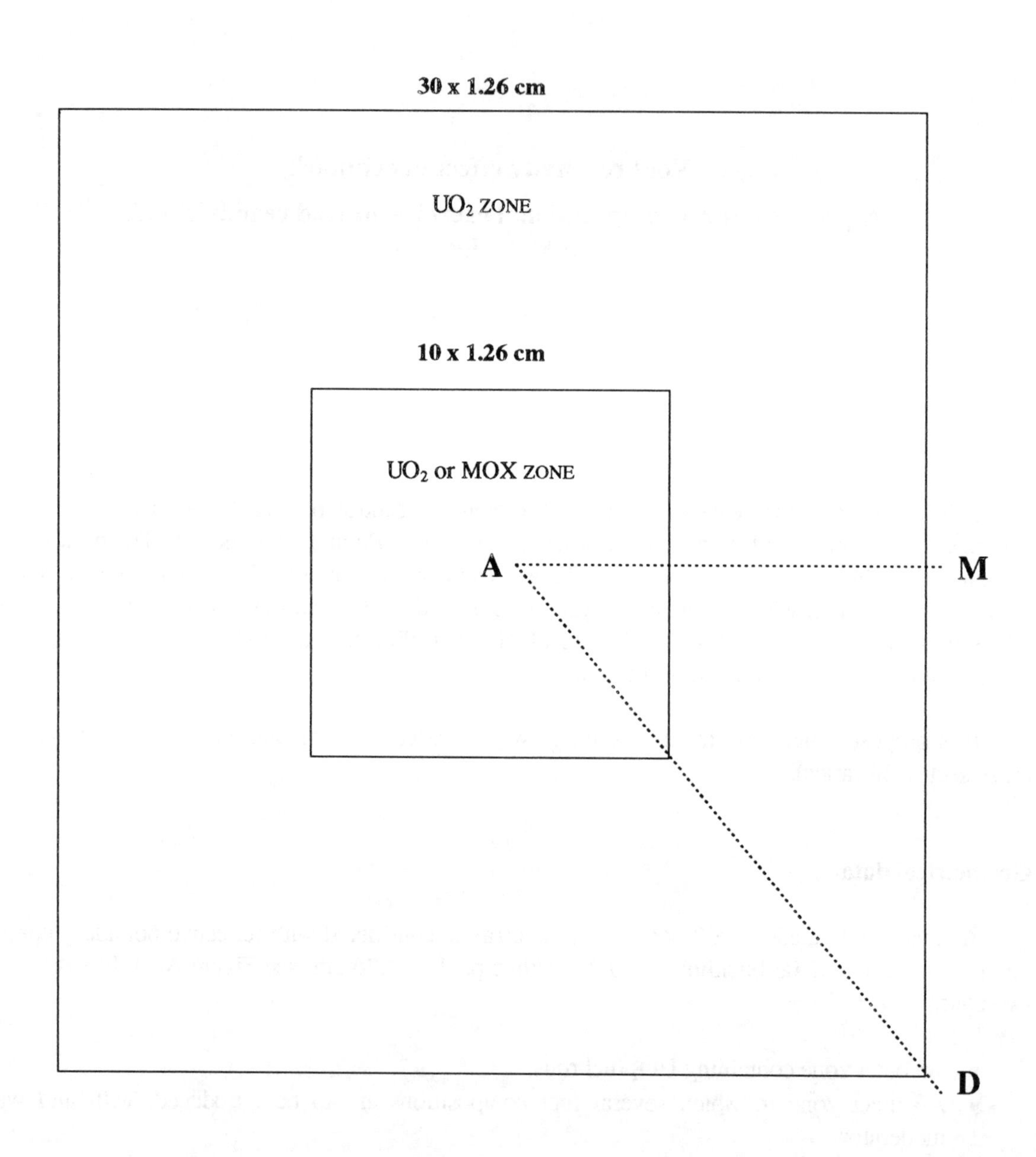

Figure A-1 **Void coefficent benchmark**

The moderator is a mixture of water and aluminium, simulating a reduced water density (in the range 0.70-0.75 as in the operating conditions of a power reactor).

Number densities (at/10^{-24} cm^3)

Four types of fuel are to be put in the inner zone: UO_2 (same as the outer zone) and MOX with three different Pu contents.

	UO_2	H-MOX	M-MOX	L-MOX
U-235	7.65E-4	7.90E-5	6.34E-5	5.67E-5
U-238	2.21E-2	1.92E-2	2.03E-2	2.14E-2
Pu-238		4.09E-5	2.95E-5	1.39E-5
Pu-239		1.99E-3	1.34E-3	7.76E-4
Pu-240		7.53E-4	5.26E-4	2.89E-4
Pu-241		2.06E-4	1.35E-4	6.61E-5
Pu-242		1.40E-4	1.03E-4	5.07E-5
Am-241		9.83E-5	6.65E-5	3.56E-5
O	4.57E-2	4.50E-2	4.51E-2	4.54E-2

	CLADDING + GAP	MODERATOR
Zr	3.68E-2	
Fe	7.23E-5	
O		2.47E-2
H		4.94E-2
Al		3.31E-3

Requested results

Results should be provided both on paper and a computer-processable medium. A short report should be provided describing:

- The calculational method used, including the name of the program and precise version;
- The data libraries used and evaluated data file from which they were derived;

- The list of isotopes for which resonance self-shielding was applied with the formalism used;
- The results should be presented, where useful, both in tabular and graphical form.

The following results should be provided in the specified order:

For each of the four fuel compositions of the inner zone:

- k-infinity of the unit cell with and without moderator (infinite lattice);
- k-infinity of the macrocell, with and without moderator in the inner zone (the outer zone is always flooded);
- Relative fission rate distributions along the fuel row nearest to the median axis (line AM on Figure A-1) and along the diagonal (line AD on Figure A-1), normalised to unity at the centre of the loading, with and without moderator in the inner zone;
- Multigroup neutron spectra per unit lethargy in the centre, in the corner cell and in the cell at the border of zone 1 on the axis AM and the energy group structure.

Appendix B.1

VIM Monte Carlo calculations

R. N. Blomquist (Argonne National Laboratory - ANL)

Generally, the continuous-energy cross-section data used was based on ENDF/B-V data. Zircaloy, however, was based on ENDF/B-IV. Specular reflection was applied at all boundary surfaces. The uncertainties shown in these results are one standard deviation of the mean. The flux and fission data submitted is normalised to per fission source neutron, and volume-integrated over each cell. Isotopic multigroup cross-sections and reaction rates are available on request.

MOX lattice void reactivity effect benchmark

Cell calculations consisted of 0.5 million histories each in generations of 10 000, with tally data grouped for each generation. Two initial generations were discarded to allow the flat fission source guess to decay. Computation time averaged 10 SPARC-2 CPU-hours.

Lattice calculations consisted of 20 million histories each in generations of 10 000, with tally data grouped for each ten generations. Five initial generations were discarded to allow the flat fission source guess to decay. Computation time averaged 5 SPARC-2 CPU-hours. Results in symmetrically-placed cells were combined to minimise the statistical uncertainties. The location of selected cells for which more detailed results are included is shown in the attached table .

A second identical set of lattice calculations, but with 22 groups, were performed for the high-Pu case, to allow for more detailed spectral comparisons.

```
The 4-group structure is as follows:
1   20 MeV - 821   KeV
2  821 KeV - 5.53  KeV
3 5.53 KeV - 0.625 eV
4 0.625 eV - 1.e-5 eV


High-Pu Unvoided Case
k-inf (infinite lattice) = 1.2124 +/- 0.0012
k-inf (macrocell)        = 1.3428 +/- 0.0002

flux densities, per source neutron, per unit lethargy
 (Emax=20MeV)

"asymptotic" voided cell spectrum (at A)
group          flux
   1  9.7024E-06 +/- 0.21%)
   2  8.2210E-06 +/- 0.19%)
   3  3.0960E-06 +/- 0.19%)
   4  2.1912E-07 +/- 0.55%)

voided region edge cell spectrum just inside voided region on AM
group          flux
   1  1.0119E-05 +/- 0.14%)
   2  8.3598E-06 +/- 0.14%)
   3  3.2381E-06 +/- 0.16%)
   4  4.2062E-07 +/- 0.23%)

uvoided region edge cell spectrum just outside voided region on AM
group          flux
   1  9.5430E-06 +/- 0.13%)
   2  8.2600E-06 +/- 0.13%)
   3  3.4683E-06 +/- 0.12%)
   4  9.1503E-07 +/- 0.16%)

voided region corner cell spectrum just inside voided region on AD
group          flux
   1  1.0342E-05 +/- 0.12%)
   2  8.4742E-06 +/- 0.10%)
   3  3.3371E-06 +/- 0.13%)
   4  5.9144E-07 +/- 0.27%)

uvoided region corner cell spectrumjust outside voided region on AD
group          flux
   1  9.5512E-06 +/- 0.12%)
   2  8.3249E-06 +/- 0.12%)
   3  3.5991E-06 +/- 0.13%)
   4  1.2224E-06 +/- 0.25%)

"asymptotic" uvoided cell spectrum (at D)
group          flux
   1  9.8750E-06 +/- 0.18%)
   2  8.5982E-06 +/- 0.20%)
   3  3.7655E-06 +/- 0.18%)
   4  1.5065E-06 +/- 0.25%)


High-Pu Voided Case
k-inf (infinite lattice) = 1.2850 +/- 0.0007
k-inf (macrocell)        = 1.3481 +/- 0.0002

flux densities, per source neutron, per unit lethargy
 (Emax=20MeV)

"asymptotic" voided cell spectrum (at A)
group          flux
   1  1.1176E-05 +/- 0.22%)
   2  1.1416E-05 +/- 0.17%)
   3  2.3238E-06 +/- 0.23%)
   4  2.9076E-08 +/- 1.78%)

voided region edge cell spectrum just inside voided region on AM
group          flux
   1  1.1201E-05 +/- 0.14%)
   2  1.0880E-05 +/- 0.08%)
   3  2.8330E-06 +/- 0.17%)
   4  2.0581E-07 +/- 0.47%)

uvoided region edge cell spectrum just outside voided region on AM
group          flux
   1  1.0288E-05 +/- 0.10%)
   2  1.0270E-05 +/- 0.07%)
   3  3.3981E-06 +/- 0.13%)
   4  6.6148E-07 +/- 0.28%)

voided region corner cell spectrum just inside voided region on AD
group          flux
   1  1.1281E-05 +/- 0.12%)
   2  1.0508E-05 +/- 0.13%)
   3  3.1918E-06 +/- 0.19%)
   4  4.0355E-07 +/- 0.26%)

uvoided region corner cell spectrumjust outside voided region on AD
group          flux
   1  1.0245E-05 +/- 0.13%)
   2  9.7326E-06 +/- 0.10%)
   3  3.8189E-06 +/- 0.16%)
   4  1.1386E-06 +/- 0.23%)

"asymptotic" uvoided cell spectrum (at D)
group          flux
   1  1.0665E-05 +/- 0.20%)
   2  9.3106E-06 +/- 0.20%)
   3  4.0867E-06 +/- 0.17%)
   4  1.6495E-06 +/- 0.18%)


Low-Pu Unvoided Case
k-inf (infinite lattice) = 1.1428 +/- 0.0011
k-inf (macrocell)        = 1.3382 +/- 0.0002

flux densities, per source neutron, per unit lethargy
 (Emax=20MeV)

"asymptotic" voided cell spectrum (at A)
group          flux
   1  9.2275E-06 +/- 0.17%)
   2  8.0005E-06 +/- 0.22%)
   3  3.3635E-06 +/- 0.24%)
   4  4.8319E-07 +/- 0.35%)

voided region edge cell spectrum just inside voided region on AM
group          flux
   1  9.6073E-06 +/- 0.09%)
   2  8.1989E-06 +/- 0.12%)
   3  3.4598E-06 +/- 0.11%)
   4  7.0349E-07 +/- 0.20%)

uvoided region edge cell spectrum just outside voided region on AM
group          flux
   1  9.3667E-06 +/- 0.13%)
   2  8.1847E-06 +/- 0.10%)
   3  3.5682E-06 +/- 0.14%)
   4  1.0797E-06 +/- 0.18%)

voided region corner cell spectrum just inside voided region on AD
```

```
group        flux
   1  9.8359E-06 +/- 0.12%)
   2  8.3167E-06 +/- 0.12%)
   3  3.5078E-06 +/- 0.11%)
   4  8.7960E-07 +/- 0.26%)

uvoided region corner cell spectrumjust outside voided region on AD
group        flux
   1  9.5298E-06 +/- 0.17%)
   2  8.3142E-06 +/- 0.12%)
   3  3.6354E-06 +/- 0.11%)
   4  1.3086E-06 +/- 0.18%)

"asymptotic" uvoided cell spectrum (at D)
group        flux
   1  9.9625E-06 +/- 0.22%)
   2  8.6656E-06 +/- 0.18%)
   3  3.7874E-06 +/- 0.20%)
   4  1.5178E-06 +/- 0.35%)

Low-Pu, Voided Case
k-inf (infinite lattice) - 0.7616 +/- 0.0006
k-inf (macrocell)        - 1.3398 +/- 0.0002

flux densities, per source neutron, per unit lethargy
 (Emax-20MeV)

"asymptotic" voided cell spectrum (at A)
group        flux
   1  1.0406E-05 +/- 0.19%)
   2  1.1200E-05 +/- 0.19%)
   3  2.7888E-06 +/- 0.31%)
   4  9.8303E-08 +/- 1.26%)

voided region edge cell spectrum just inside voided region on AM
group        flux
   1  1.0404E-05 +/- 0.15%)
   2  1.0639E-05 +/- 0.11%)
   3  3.1777E-06 +/- 0.16%)
   4  3.1754E-07 +/- 0.32%)

uvoided region edge cell spectrum just outside voided region on AM
group        flux
   1  9.8706E-06 +/- 0.16%)
   2  1.0109E-05 +/- 0.08%)
   3  3.5953E-06 +/- 0.12%)
   4  7.4991E-07 +/- 0.32%)

voided region corner cell spectrum just inside voided region on AD
group        flux
   1  1.0588E-05 +/- 0.16%)
   2  1.0295E-05 +/- 0.10%)
   3  3.4566E-06 +/- 0.17%)
   4  5.4513E-07 +/- 0.29%)

uvoided region corner cell spectrumjust outside voided region on AD
group        flux
   1  1.0060E-05 +/- 0.14%)
   2  9.6583E-06 +/- 0.09%)
   3  3.9102E-06 +/- 0.15%)
   4  1.2011E-06 +/- 0.30%)

"asymptotic" uvoided cell spectrum (at D)
group        flux
   1  1.0860E-05 +/- 0.22%)
   2  9.4127E-06 +/- 0.16%)
   3  4.1366E-06 +/- 0.19%)
   4  1.6614E-06 +/- 0.26%)
```



```
Mid-Pu, Unvoided Case
k-inf (infinite lattice) - 1.1671 +/- 0.0011
k-inf (macrocell)        - 1.3391 +/- 0.0002

flux densities, per source neutron, per unit lethargy
 (Emax-20MeV)

"asymptotic" voided cell spectrum (at A)
group        flux
   1  9.4467E-06 +/- 0.16%)
   2  8.0918E-06 +/- 0.17%)
   3  3.2109E-06 +/- 0.19%)
   4  2.9744E-07 +/- 0.61%)

voided region edge cell spectrum just inside voided region on AM
group        flux
   1  9.8507E-06 +/- 0.13%)
   2  8.2679E-06 +/- 0.15%)
   3  3.3219E-06 +/- 0.13%)
   4  5.1242E-07 +/- 0.15%)

uvoided region edge cell spectrum just outside voided region on AM
group        flux
   1  9.4108E-06 +/- 0.17%)
   2  8.1961E-06 +/- 0.10%)
   3  3.5083E-06 +/- 0.15%)
   4  9.6637E-07 +/- 0.28%)

voided region corner cell spectrum just inside voided region on AD
group        flux
   1  1.0112E-05 +/- 0.14%)
   2  8.3872E-06 +/- 0.14%)
   3  3.4164E-06 +/- 0.15%)
   4  6.9115E-07 +/- 0.40%)
uvoided region corner cell spectrumjust outside voided region on AD
group        flux
   1  9.5619E-06 +/- 0.17%)
   2  8.3047E-06 +/- 0.12%)
   3  3.6199E-06 +/- 0.15%)
   4  1.2550E-06 +/- 0.27%)

"asymptotic" uvoided cell spectrum (at D)
group        flux
   1  9.9537E-06 +/- 0.18%)
   2  8.6486E-06 +/- 0.19%)
   3  3.7729E-06 +/- 0.19%)
   4  1.5130E-06 +/- 0.40%)

Mid-Pu Voided Case
k-inf (infinite lattice) - 1.0380 +/- 0.0006
k-inf (macrocell)        - 1.3434 +/- 0.0002

flux densities, per source neutron, per unit lethargy
 (Emax-20MeV)

"asymptotic" voided cell spectrum (at A)
group        flux
   1  1.0768E-05 +/- 0.20%)
   2  1.1299E-05 +/- 0.20%)
   3  2.5266E-06 +/- 0.19%)
   4  4.7225E-08 +/- 1.36%)

voided region edge cell spectrum just inside voided region on AM
group        flux
   1  1.0804E-05 +/- 0.11%)
   2  1.0764E-05 +/- 0.10%)
   3  2.9822E-06 +/- 0.16%)
```

```
   4  2.4474E-07 +/- 0.35%)

uvoided region edge cell spectrum just outside voided region on AM
group        flux
   1  1.0089E-05 +/- 0.11%)
   2  1.0192E-05 +/- 0.08%)
   3  3.4882E-06 +/- 0.14%)
   4  6.9656E-07 +/- 0.23%)

voided region corner cell spectrum just inside voided region on AD
group        flux
   1  1.0950E-05 +/- 0.11%)
   2  1.0412E-05 +/- 0.07%)
   3  3.2955E-06 +/- 0.13%)
   4  4.5351E-07 +/- 0.41%)

uvoided region corner cell spectrumjust outside voided region on AD
group        flux
   1  1.0170E-05 +/- 0.11%)
   2  9.7081E-06 +/- 0.11%)
   3  3.8500E-06 +/- 0.10%)
   4  1.1545E-06 +/- 0.26%)

"asymptotic" uvoided cell spectrum (at D)
group        flux
   1  1.0790E-05 +/- 0.18%)
   2  9.4007E-06 +/- 0.17%)
   3  4.1174E-06 +/- 0.16%)
   4  1.6499E-06 +/- 0.16%)

UOX, Unvoided Case
k-inf (infinite lattice) - 1.3651 +/- 0.0010
k-inf (macrocell)        - 1.3653 +/- 0.0002

flux densities, per source neutron, per unit lethargy
 (Emax-20MeV)

"asymptotic" voided cell spectrum (at A)
group        flux
   1  9.7036E-06 +/- 0.17%)
   2  8.4622E-06 +/- 0.20%)
   3  3.7162E-06 +/- 0.18%)
   4  1.4925E-06 +/- 0.28%)

voided region edge cell spectrum just inside voided region on AM
group        flux
   1  9.7279E-06 +/- 0.09%)
   2  8.4612E-06 +/- 0.11%)
   3  3.7089E-06 +/- 0.10%)
   4  1.4944E-06 +/- 0.20%)

uvoided region edge cell spectrum just outside voided region on AM
group        flux
   1  9.7059E-06 +/- 0.12%)
   2  8.4631E-06 +/- 0.11%)
   3  3.7127E-06 +/- 0.11%)
   4  1.4957E-06 +/- 0.15%)

voided region corner cell spectrum just inside voided region on AD
group        flux
   1  9.7112E-06 +/- 0.14%)
   2  8.4754E-06 +/- 0.09%)
   3  3.7120E-06 +/- 0.16%)
   4  1.4984E-06 +/- 0.23%)

uvoided region corner cell spectrumjust outside voided region on AD
group        flux
   1  9.7162E-06 +/- 0.15%)
   2  8.4628E-06 +/- 0.10%)
   3  3.7194E-06 +/- 0.17%)
   4  1.4918E-06 +/- 0.34%)

"asymptotic" uvoided cell spectrum (at D)
group        flux
   1  9.7125E-06 +/- 0.24%)
   2  8.4622E-06 +/- 0.17%)
   3  3.7110E-06 +/- 0.15%)
   4  1.4871E-06 +/- 0.38%)

UOX, Voided Case
k-inf (infinite lattice) - 0.6215 +/- 0.0006
k-inf (macrocell)        - 1.3508 +/- 0.0002

flux densities, per source neutron, per unit lethargy
 (Emax-20MeV)

"asymptotic" voided cell spectrum (at A)
group        flux
   1  1.0488E-05 +/- 0.21%)
   2  1.1344E-05 +/- 0.20%)
   3  3.1441E-06 +/- 0.22%)
   4  4.4127E-07 +/- 0.27%)

voided region edge cell spectrum just inside voided region on AM
group        flux
   1  1.0299E-05 +/- 0.12%)
   2  1.0737E-05 +/- 0.09%)
   3  3.4554E-06 +/- 0.11%)
   4  6.5147E-07 +/- 0.27%)

uvoided region edge cell spectrum just outside voided region on AM
group        flux
   1  9.9465E-06 +/- 0.12%)
   2  1.0224E-05 +/- 0.13%)
   3  3.7464E-06 +/- 0.14%)
   4  9.8129E-07 +/- 0.25%)

voided region corner cell spectrum just inside voided region on AD
group        flux
   1  1.0283E-05 +/- 0.10%)
   2  1.0322E-05 +/- 0.13%)
   3  3.6649E-06 +/- 0.13%)
   4  8.6228E-07 +/- 0.25%)

uvoided region corner cell spectrumjust outside voided region on AD
group        flux
   1  1.0100E-05 +/- 0.14%)
   2  9.7251E-06 +/- 0.13%)
   3  3.9774E-06 +/- 0.14%)
   4  1.3277E-06 +/- 0.24%)

"asymptotic" uvoided cell spectrum (at D)
group        flux
   1  1.0711E-05 +/- 0.16%)
   2  9.3812E-06 +/- 0.22%)
   3  4.1070E-06 +/- 0.24%)
   4  1.6495E-06 +/- 0.26%)
```

Appendix B.2

Benchmarks on physics of plutonium recycling

M. Soldevila (CEA) and M. Aigle (Framatome)

Void reactivity effect benchmark

Code used for all calculations:	**APOLLO-2**
Library:	**CEA 93** with 172 energy groups in *cell* calculations 99 energy groups in *macrocell* calculations
Evaluated data file:	**JEF-2.2** for main isotopes
Self-shielding isotopes:	U-235, U-238, Pu-238, Pu-239, Pu-240, Pu-241, Pu-242, Am-241, Zr
Self-shielding formalism for spatial interaction:	Background matrix with 6 rings in the fuel region

Flux calculation by collision probability method.

More details are provided in Puill's and Kolmayer's contribution.

Appendix B.3

Results of the void reactivity effect benchmark in MOX fuelled PWRs

G. Rimpault and S. Rahlfs (CEA)

The unit cell calculations have been carried out with the European Cell Code ECCO version 5.2 in 1968 energy groups.

The P1 consistent Flux solution method with the collision probabilities for the heterogeneous treatment and the subgroup method for the resonance self-shielding has been used.

The data library was the JEF-2/ECCO library in 1968 energy groups. Pu-238, only available in 172 energy groups was deconvoluted in ECCO into 1968 groups.

	RESONANCE SELF-SHIELDING	EVALUATED DATA FILES
U-235	YES	JEF-2
U-238	YES	JEF-2
Pu-238	YES	JEF-2
Pu-239	YES	JEF-2
Pu-240	YES	JEF-2
Pu-241	YES	JEF-2
Pu-242	YES	JEF-2
Am-241	YES	JEF-2
O	YES	JEF-2
Zr	YES	JEF-2
Fe-54	YES	JEF-2
Fe-56	YES	JEF-2
Fe-57	YES	JEF-2
Fe-58	YES	JEF-2
H	YES	JEF-2
Al	YES	JEF-2

The macrocell calculations have been carried out with the ERANOS code version 1.0.

Homogenised P1 cross-sections in 172 groups from the ECCO calculations have been used for the S_n transport calculations in ERANOS.

Appendix B.4

APOLLO-1 calculations

S. Cathalau and A. Maghnouj (CEA)

Generalities

The APOLLO-1 code solves the Boltzmann Equation (integral form) in space and energy, in a multigroup approximation using a collision probability method.

The CEA-86.1 library (99 groups, 52 fast and 47 thermal with a cut-off at 2.77 eV, E0 = 10 MeV) is based essentially on the JEF-1 and ENDF/B-V evaluations and includes also internal evaluations:

- U-235 ENDF/B-V including the η variation in thermal range,
- U-238 ENDF/B-V including a fictitious negative resonance near the 0 eV,
- Pu-239 JEF-1 including the ν variation in the first resonance,
- Pu-240 ENDF/B-IV + RIBON evaluation,
- Pu-241 internal evaluation,
- Pu-242 ENDF/B-V,
- Am-241 Internal evaluation.

Self-shielding calculations have been performed using the Livollant-Jeanpierre formalism and specific modules are implemented in the code in order to take into account resonance overlapping effect and self-shielding and Doppler broadening effects for Pu-240 and Pu-242 "thermal" resonances (1.054 eV and 2.68 eV respectively).

Only U-235, U-238, Pu-239, Pu-240, Pu-242, and natural Zr have been self-shielded.

Cell infinite calculations

All calculations have been performed assuming a cylindrization for the cell.

Appendix B.5

Void reactivity effect benchmark
Computational exercise on void reactivity effect in MOX

P. A. Landeyro (ENEA)

Calculation method

All the calculations were carried out with the MCNP Monte Carlo code, version 4.2.

The cross-sections of Al, Fe and natural Zr were taken from the ENDL-85 library, that is included in the MCNP package. The cross sections of Am-241 were produced at ENEA from the Joint Evaluated File (JEF), version 2.2. All the cross-sections corresponding to the rest of the nuclides were determined from the JEF-1 file.

Appendix B.6

Void reactivity effect benchmark

D. C. Lutz (IKE – University of Stuttgart)

The cross-section basis is the JEF-1 data file [1], which is processed [2] with the NJOY program system [3] into multigroup data. Three libraries are available in the following energy ranges:

- Fast and epithermal range, 100 groups,
- The resolved resonance region, 8500 groups,
- The thermal range, 151 groups.

These libraries are used in the code CGM [4] for one-dimensional, usually 5 zones spectral calculations for group collapsing to 60 (25 fast, 35 thermal) groups. The generated data sets are dependent on the cell definition in CGM, mainly in the groups containing resonances. They include the effects of self and mutual shielding and resonance overlapping for the U- and Pu-isotopes. In standard calculations only 1 set of cross-sections is produced for the whole life of fuel, but in this case the case the cross-sections have been recalculated 4 times during the burnup.

The pin cell calculations for the homogenisation spectra are carried out in RSYST 3 [6] for a cylindrical cell using a collision probability code.

The assembly calculations are also performed in RSYST with the J±-code ICM2D [6]. The code takes into account a linear space and angle-dependent flux distribution in the cell and has an option for treating heterogeneous substructures of complicated geometry by imbedded collision probability (P_{ik}) calculations.

References

[1] J. L. Rowlands, N. Tubbs, "The Joint Evaluated File: A New Data Library for Reactor Calculations", Intl. Conf. on Nuclear data for Basic Applied Science, Santa Fe, U.S.A., 1985.

[2] M. Mattes, IKE, Internal report, 1989.

[3] R. E. Mac Farlane et al., "The NJOY Nuclear Data Processing System", LA-9309-M, 1985.

[4] M. Arshad, "Development and Validation of a Program System for Calculation of Spectra and Weighted Group Constants for Thermal and Epithermal Systems", Thesis University of Stuttgart, (IKE 6-156), 1986.

[5] R. Rühle, "RSYST An Integrated Modular System for Reactor and Shielding Calculations", USAEC Conf-730 414-12, 1973.

[6] W. Bernnat, D. Emendörfer, D. Lutz, Th. Ruckle and B. Szczesna, "Physical and Computational Properties of a Multigroup Assembly Code Based on the J± and P_{ik} concept", ENC, Paris, 1987.

Appendix B.7

Void reactivity effect benchmark

M. Mattes, W. Bernnat and S. Käfer (IKE – University of Stuttgart)

The calculations were performed with the general-purpose, continuous-energy, generalised-geometry Monte Carlo transport code MCNP, version 4.2.

The cross-section libraries used were produced from the JEF-2.2 evaluated files with the nuclear data processing code NJOY-91.91 with some modifications. The resonance reconstruction in RECONR was done with a 0.1% accuracy and no thinning was applied to the pointwise cross-section data in ACER.

As MCNP 4.2 cannot treat double-differential data given in file MF = 6 (energy-angle correlation), such representations in the evaluated files were transformed into MF = 4 and MF = 5, for outgoing neutrons, with the code SIXPAK before processing with NJOY.

In the case of this benchmark the only reactions and nuclide affected by this procedure were (n,2n), (n,3n), (n,4n) and (n,n' continuum) of U-235.

Each of the assembly calculations run with 5 million neutron histories. Neutron flux spectra are given for 292 energy groups from 0.00001 eV up to 20 MeV.

Appendix B.8

Void reactivity effect benchmark

H. Takano and H. Akie (JAERI) and K. Kaneko (Japan Research Institute)

The benchmark of void reactivity effect in MOX lattices was performed with the SRAC system and the MVP code.

Description of the SRAC calculation

- *Calculational method, program name and version*: collision probability method for an infinite lattice with 107 energy groups, and diffusion calculation for the macrocell with 107 groups. The modified version of SRAC [1] was used;

- *Data libraries and original evaluated data file*: SRACLIB-JENDL3 processed from JENDL-3.1 [2] data file.

- *List of isotopes for resonance shielding calculation and method used*: the self-shielding factor table (f-table) interpolation method can be applied in the whole energy region. In the resolved resonance region (E < 961 eV in the SRAC system), PEACO [3] can treat both the self-shielding and the mutual resonance overlapping effects by the ultra-fine group method, which calculates the spectrum with the energy structure of lethargy width Δu = 2.5E-4 between 961 eV and 130 eV and Δu = 5.0E-4 between 130 eV and 2.38 eV (2.38 eV is the thermal cut-off energy in the calculations).

The PEACO method was used for U-235, U-238, Pu-238, Pu-239, Pu-240, Pu-241 and Pu-242. The self-shielding effect in the thermal range (E < 2.38 eV) was considered with the f-table method for U-235, Pu-238, Pu-239, Pu-240 and Pu-242. For other nuclides, the f-table method was used in the fast energy range.

Description of the MVP calculation

- *Calculational method, program name and version*: continuous-energy Monte Carlo calculation was made with the MVP [4] code.

- *Data libraries and original evaluated data file*: the library was processed from JENDL-3.1 data file.

- *List of isotopes for resonance shielding calculation and method used*: continuous-energy Monte Carlo method with the pointwise cross-section data for all nuclides. The probability table method was adopted for the unresolved resonances.

References

[1] K. Tsuchihashi et al., JAERI-1302 (1986).

[2] K. Shibata et al., JAERI-1319 (1990).

[3] Y. Ishiguro, JAERI-M-5527 (1974).

[4] T. Mori et al., J. Nucl. Sci. Technol., 29, 325 (1992).

```
## NEACRP PWR VOID COEFFICIENT BENCHMARK BY SRAC-DIFFUSION ## UO2 UNVOIDED

  FLUX DENSITY , PER SOURCE NETUTRON , PER LETHARGY

 << VOIDED REGION CENTRAL CELL SPECTRUM >>
  GROUP    ELOWER(EV)       FLUX
     1    8.20850E+05    3.01634E-03
     2    5.53080E+03    2.03274E-03
     3    6.02360E-01    8.96517E-04
     4    1.00000E-05    3.57006E-04

 << VOIDED REGION EDGE CELL SPECTRUM JUST INSIDE VOIDED REGION ON AM >>
  GROUP    ELOWER(EV)       FLUX
     1    8.20850E+05    2.99725E-03
     2    5.53080E+03    2.03289E-03
     3    6.02360E-01    8.96506E-04
     4    1.00000E-05    3.57024E-04

 << UNVOIDED REGION EDGE CELL SPECTRUM JUST OUTSIDE VOIDED REGION ON AM >>
  GROUP    ELOWER(EV)       FLUX
     1    8.20850E+05    2.99715E-03
     2    5.53080E+03    2.03297E-03
     3    6.02360E-01    8.96488E-04
     4    1.00000E-05    3.57037E-04

 << VOIDED REGION EDGE CELL SPECTRUM JUST INSIDE VOIDED REGION ON AD >>
  GROUP    ELOWER(EV)       FLUX
     1    8.20850E+05    2.99712E-03
     2    5.53080E+03    2.03296E-03
     3    6.02360E-01    8.96497E-04
     4    1.00000E-05    3.57035E-04

 << UNVOIDED REGION EDGE CELL SPECTRUM JUST OUTSIDE VOIDED REGION ON AD >>
  GROUP    ELOWER(EV)       FLUX
     1    8.20850E+05    2.99699E-03
     2    5.53080E+03    2.03304E-03
     3    6.02360E-01    8.96480E-04
     4    1.00000E-05    3.57048E-04

 << UNVOIDED CELL SPECTRUM (AT D) >>
  GROUP    ELOWER(EV)       FLUX
     1    8.20850E+05    2.99680E-03
     2    5.53080E+03    2.03305E-03
     3    6.02360E-01    8.96490E-04
     4    1.00000E-05    3.57059E-04


## NEACRP PWR VOID COEFFICIENT BENCHMARK BY SRAC-DIFFUSION ## UO2 VOIDED

  FLUX DENSITY , PER SOURCE NETUTRON , PER LETHARGY

 << VOIDED REGION CENTRAL CELL SPECTRUM >>
  GROUP    ELOWER(EV)       FLUX
     1    8.20850E+05    3.21357E-03
     2    5.53080E+03    2.68242E-03
     3    6.02360E-01    7.97361E-04
     4    1.00000E-05    1.03105E-04

 << VOIDED REGION EDGE CELL SPECTRUM JUST INSIDE VOIDED REGION ON AM >>
  GROUP    ELOWER(EV)       FLUX
     1    8.20850E+05    3.16140E-03
     2    5.53080E+03    2.56199E-03
     3    6.02360E-01    8.62401E-04
     4    1.00000E-05    1.53982E-04

 << UNVOIDED REGION EDGE CELL SPECTRUM JUST OUTSIDE VOIDED REGION ON AM >>
  GROUP    ELOWER(EV)       FLUX
     1    8.20850E+05    3.12593E-03
     2    5.53080E+03    2.47479E-03
     3    6.02360E-01    9.14483E-04
     4    1.00000E-05    2.28358E-04

 << VOIDED REGION EDGE CELL SPECTRUM JUST INSIDE VOIDED REGION ON AD >>
  GROUP    ELOWER(EV)       FLUX
     1    8.20850E+05    3.15594E-03
     2    5.53080E+03    2.46482E-03
     3    6.02360E-01    9.09284E-04
     4    1.00000E-05    2.02347E-04

 << UNVOIDED REGION EDGE CELL SPECTRUM JUST OUTSIDE VOIDED REGION ON AD >>
  GROUP    ELOWER(EV)       FLUX
     1    8.20850E+05    3.14473E-03
     2    5.53080E+03    2.34802E-03
     3    6.02360E-01    9.68535E-04
     4    1.00000E-05    3.14099E-04

 << UNVOIDED CELL SPECTRUM (AT D) >>
  GROUP    ELOWER(EV)       FLUX
     1    8.20850E+05    3.30601E-03
     2    5.53080E+03    2.24529E-03
     3    6.02360E-01    9.89270E-04
     4    1.00000E-05    3.94103E-04


## NEACRP PWR VOID COEFFICIENT BENCHMARK BY SRAC-DIFFUSION ## H-MOX UNVOIDED


   RESULT BY SRAC-DIFF.: UPPER ENERGY LIMIT  IS 10000000. EV
                       : No. OF ENERGY GROUPS IS 107G(LATTICE)/107G(MACROCELL)
                       : NO AXIAL BUCKLING, DIFFUISON COEFF. IS ANISOTROPIC

  FLUX DENSITY , PER SOURCE NETUTRON , PER LETHARGY

 << VOIDED REGION CENTRAL CELL SPECTRUM >>
  GROUP    ELOWER(EV)       FLUX
     1    8.20850E+05    3.05688E-03
     2    5.53080E+03    1.99285E-03
     3    6.02360E-01    7.51636E-04
     4    1.00000E-05    5.39031E-05

 << VOIDED REGION EDGE CELL SPECTRUM JUST INSIDE VOIDED REGION ON AM >>
  GROUP    ELOWER(EV)       FLUX
     1    8.20850E+05    3.04662E-03
     2    5.53080E+03    2.00802E-03
     3    6.02360E-01    7.93295E-04
     4    1.00000E-05    1.09476E-04

 << UNVOIDED REGION EDGE CELL SPECTRUM JUST OUTSIDE VOIDED REGION ON AM >>
  GROUP    ELOWER(EV)       FLUX
     1    8.20850E+05    2.99701E-03
     2    5.53080E+03    2.00008E-03
     3    6.02360E-01    8.35059E-04
     4    1.00000E-05    2.11592E-04

 << VOIDED REGION EDGE CELL SPECTRUM JUST INSIDE VOIDED REGION ON AD >>
  GROUP    ELOWER(EV)       FLUX
     1    8.20850E+05    3.04953E-03
     2    5.53080E+03    2.01782E-03
     3    6.02360E-01    8.22100E-04
     4    1.00000E-05    1.57903E-04

 << UNVOIDED REGION EDGE CELL SPECTRUM JUST OUTSIDE VOIDED REGION ON AD >>
  GROUP    ELOWER(EV)       FLUX
     1    8.20850E+05    2.98622E-03
     2    5.53080E+03    2.00842E-03
     3    6.02360E-01    8.66753E-04
```

```
    4    1.00000E-05    2.85213E-04

 << UNVOIDED CELL SPECTRUM (AT D) >>
  GROUP    ELOWER(EV)       FLUX
     1    8.20850E+05    3.03893E-03
     2    5.53080E+03    2.05649E-03
     3    6.02360E-01    9.03995E-04
     4    1.00000E-05    3.59198E-04

## NEACRP PWR VOID COEFFICIENT BENCHMARK BY SRAC-DIFFUSION ## H-MOX VOIDED

  FLUX DENSITY , PER SOURCE NETUTRON , PER LETHARGY

 << VOIDED REGION CENTRAL CELL SPECTRUM >>
  GROUP    ELOWER(EV)       FLUX
     1    8.20850E+05    3.37867E-03
     2    5.53080E+03    2.68210E-03
     3    6.02360E-01    6.02684E-04
     4    1.00000E-05    1.44265E-06

 << VOIDED REGION EDGE CELL SPECTRUM JUST INSIDE VOIDED REGION ON AM >>
  GROUP    ELOWER(EV)       FLUX
     1    8.20850E+05    3.31303E-03
     2    5.53080E+03    2.57158E-03
     3    6.02360E-01    7.19816E-04
     4    1.00000E-05    3.69853E-05

 << UNVOIDED REGION EDGE CELL SPECTRUM JUST OUTSIDE VOIDED REGION ON AM >>
  GROUP    ELOWER(EV)       FLUX
     1    8.20850E+05    3.22338E-03
     2    5.53080E+03    2.47897E-03
     3    6.02360E-01    8.22910E-04
     4    1.00000E-05    1.38416E-04

 << VOIDED REGION EDGE CELL SPECTRUM JUST INSIDE VOIDED REGION ON AD >>
  GROUP    ELOWER(EV)       FLUX
     1    8.20850E+05    3.29179E-03
     2    5.53080E+03    2.47857E-03
     3    6.02360E-01    8.05162E-04
     4    1.00000E-05    7.96489E-05

 << UNVOIDED REGION EDGE CELL SPECTRUM JUST OUTSIDE VOIDED REGION ON AD >>
  GROUP    ELOWER(EV)       FLUX
     1    8.20850E+05    3.18624E-03
     2    5.53080E+03    2.34557E-03
     3    6.02360E-01    9.24968E-04
     4    1.00000E-05    2.52812E-04

 << UNVOIDED CELL SPECTRUM (AT D) >>
  GROUP    ELOWER(EV)       FLUX
     1    8.20850E+05    3.29617E-03
     2    5.53080E+03    2.23625E-03
     3    6.02360E-01    9.84566E-04
     4    1.00000E-05    3.91988E-04


## NEACRP PWR VOID COEFFICIENT BENCHMARK BY SRAC-DIFFUSION ## M-MOX UNVOIDED

  FLUX DENSITY , PER SOURCE NETUTRON , PER LETHARGY

 << VOIDED REGION CENTRAL CELL SPECTRUM >>
  GROUP    ELOWER(EV)       FLUX
     1    8.20850E+05    2.98748E-03
     2    5.53080E+03    1.96798E-03
     3    6.02360E-01    7.82958E-04
     4    1.00000E-05    7.33313E-05

 << VOIDED REGION EDGE CELL SPECTRUM JUST INSIDE VOIDED REGION ON AM >>
  GROUP    ELOWER(EV)       FLUX
     1    8.20850E+05    2.99617E-03
     2    5.53080E+03    1.99106E-03
     3    6.02360E-01    8.15504E-04
     4    1.00000E-05    1.32158E-04

 << UNVOIDED REGION EDGE CELL SPECTRUM JUST OUTSIDE VOIDED REGION ON AM >>
  GROUP    ELOWER(EV)       FLUX
     1    8.20850E+05    2.96470E-03
     2    5.53080E+03    1.98883E-03
     3    6.02360E-01    8.47497E-04
     4    1.00000E-05    2.25885E-04

 << VOIDED REGION EDGE CELL SPECTRUM JUST INSIDE VOIDED REGION ON AD >>
  GROUP    ELOWER(EV)       FLUX
     1    8.20850E+05    3.01395E-03
     2    5.53080E+03    2.00727E-03
     3    6.02360E-01    8.38444E-04
     4    1.00000E-05    1.81009E-04

 << UNVOIDED REGION EDGE CELL SPECTRUM JUST OUTSIDE VOIDED REGION ON AD >>
  GROUP    ELOWER(EV)       FLUX
     1    8.20850E+05    2.97609E-03
     2    5.53080E+03    2.00624E-03
     3    6.02360E-01    8.73154E-04
     4    1.00000E-05    2.94428E-04

 << UNVOIDED CELL SPECTRUM (AT D) >>
  GROUP    ELOWER(EV)       FLUX
     1    8.20850E+05    3.05970E-03
     2    5.53080E+03    2.07034E-03
     3    6.02360E-01    9.09599E-04
     4    1.00000E-05    3.61281E-04

## NEACRP PWR VOID COEFFICIENT BENCHMARK BY SRAC-DIFFUSION ## H-MOX VOIDED

  FLUX DENSITY , PER SOURCE NETUTRON , PER LETHARGY

 << VOIDED REGION CENTRAL CELL SPECTRUM >>
  GROUP    ELOWER(EV)       FLUX
     1    8.20850E+05    3.37867E-03
     2    5.53080E+03    2.68210E-03
     3    6.02360E-01    6.02684E-04
     4    1.00000E-05    1.44265E-06

 << VOIDED REGION EDGE CELL SPECTRUM JUST INSIDE VOIDED REGION ON AM >>
  GROUP    ELOWER(EV)       FLUX
     1    8.20850E+05    3.31303E-03
     2    5.53080E+03    2.57158E-03
     3    6.02360E-01    7.19816E-04
     4    1.00000E-05    3.69853E-05

 << UNVOIDED REGION EDGE CELL SPECTRUM JUST OUTSIDE VOIDED REGION ON AM >>
  GROUP    ELOWER(EV)       FLUX
     1    8.20850E+05    3.22338E-03
     2    5.53080E+03    2.47897E-03
     3    6.02360E-01    8.22910E-04
     4    1.00000E-05    1.38416E-04

 << VOIDED REGION EDGE CELL SPECTRUM JUST INSIDE VOIDED REGION ON AD >>
  GROUP    ELOWER(EV)       FLUX
     1    8.20850E+05    3.29179E-03
     2    5.53080E+03    2.47857E-03
     3    6.02360E-01    8.05162E-04
     4    1.00000E-05    7.96489E-05

 << UNVOIDED REGION EDGE CELL SPECTRUM JUST OUTSIDE VOIDED REGION ON AD >>
  GROUP    ELOWER(EV)       FLUX
```

```
     1    8.20850E+05    3.18624E-03
     2    5.53080E+03    2.34557E-03
     3    6.02360E-01    9.24968E-04
     4    1.00000E-05    2.52812E-04

 << UNVOIDED CELL SPECTRUM (AT D) >>
  GROUP    ELOWER(EV)       FLUX
     1    8.20850E+05    3.29617E-03
     2    5.53080E+03    2.23625E-03
     3    6.02360E-01    9.84566E-04
     4    1.00000E-05    3.91988E-04


## NEACRP PWR VOID COEFFICIENT BENCHMARK BY SRAC-DIFFUSION ## L-MOX UNVOIDED

  FLUX DENSITY , PER SOURCE NETUTRON , PER LETHARGY

 << VOIDED REGION CENTRAL CELL SPECTRUM >>
  GROUP    ELOWER(EV)       FLUX
     1    8.20850E+05    2.92872E-03
     2    5.53080E+03    1.94752E-03
     3    6.02360E-01    8.21620E-04
     4    1.00000E-05    1.17873E-04

 << VOIDED REGION EDGE CELL SPECTRUM JUST INSIDE VOIDED REGION ON AM >>
  GROUP    ELOWER(EV)       FLUX
     1    8.20850E+05    2.94975E-03
     2    5.53080E+03    1.97618E-03
     3    6.02360E-01    8.42602E-04
     4    1.00000E-05    1.77547E-04

 << UNVOIDED REGION EDGE CELL SPECTRUM JUST OUTSIDE VOIDED REGION ON AM >>
  GROUP    ELOWER(EV)       FLUX
     1    8.20850E+05    2.93847E-03
     2    5.53080E+03    1.98025E-03
     3    6.02360E-01    8.62687E-04
     4    1.00000E-05    2.52840E-04

 << VOIDED REGION EDGE CELL SPECTRUM JUST INSIDE VOIDED REGION ON AD >>
  GROUP    ELOWER(EV)       FLUX
     1    8.20850E+05    2.97863E-03
     2    5.53080E+03    1.99736E-03
     3    6.02360E-01    8.58040E-04
     4    1.00000E-05    2.23231E-04

 << UNVOIDED REGION EDGE CELL SPECTRUM JUST OUTSIDE VOIDED REGION ON AD >>
  GROUP    ELOWER(EV)       FLUX
     1    8.20850E+05    2.96918E-03
     2    5.53080E+03    2.00536E-03
     3    6.02360E-01    8.80567E-04
     4    1.00000E-05    3.09686E-04

 << UNVOIDED CELL SPECTRUM (AT D) >>
  GROUP    ELOWER(EV)       FLUX
     1    8.20850E+05    3.07303E-03
     2    5.53080E+03    2.07935E-03
     3    6.02360E-01    9.13406E-04
     4    1.00000E-05    3.62740E-04


## NEACRP PWR VOID COEFFICIENT BENCHMARK BY SRAC-DIFFUSION ## L-MOX VOIDED

  FLUX DENSITY , PER SOURCE NETUTRON , PER LETHARGY

 << VOIDED REGION CENTRAL CELL SPECTRUM >>
  GROUP    ELOWER(EV)       FLUX
     1    8.20850E+05    3.20586E-03
     2    5.53080E+03    2.64802E-03
     3    6.02360E-01    7.15521E-04
     4    1.00000E-05    1.49335E-05

 << VOIDED REGION EDGE CELL SPECTRUM JUST INSIDE VOIDED REGION ON AM >>
  GROUP    ELOWER(EV)       FLUX
     1    8.20850E+05    3.16899E-03
     2    5.53080E+03    2.53703E-03
     3    6.02360E-01    8.02144E-04
     4    1.00000E-05    6.88602E-05

 << UNVOIDED REGION EDGE CELL SPECTRUM JUST OUTSIDE VOIDED REGION ON AM >>
  GROUP    ELOWER(EV)       FLUX
     1    8.20850E+05    3.11810E-03
     2    5.53080E+03    2.44928E-03
     3    6.02360E-01    8.77264E-04
     4    1.00000E-05    1.63993E-04

 << VOIDED REGION EDGE CELL SPECTRUM JUST INSIDE VOIDED REGION ON AD >>
  GROUP    ELOWER(EV)       FLUX
     1    8.20850E+05    3.17424E-03
     2    5.53080E+03    2.44795E-03
     3    6.02360E-01    8.64748E-04
     4    1.00000E-05    1.24621E-04

 << UNVOIDED REGION EDGE CELL SPECTRUM JUST OUTSIDE VOIDED REGION ON AD >>
  GROUP    ELOWER(EV)       FLUX
     1    8.20850E+05    3.13761E-03
     2    5.53080E+03    2.33093E-03
     3    6.02360E-01    9.50814E-04
     4    1.00000E-05    2.75726E-04

 << UNVOIDED CELL SPECTRUM (AT D) >>
  GROUP    ELOWER(EV)       FLUX
     1    8.20850E+05    3.33724E-03
     2    5.53080E+03    2.26334E-03
     3    6.02360E-01    9.95322E-04
     4    1.00000E-05    3.95944E-04
```

```
## NEACRP PWR VOID COEFFICIENT BENCHMARK BY MVP ## UO2 UNVOIDED

 FLUX DENSITY , PER SOURCE NETUTRON , PER LETHARGY

 << VOIDED REGION CENTRAL CELL SPECTRUM >>
 GROUP    ELOWER(EV)        FLUX
    1    1.00000E+07    2.07617E-05 +/- 15.239 %
    2    8.20850E+05    3.08131E-03 +/-  0.663 %
    3    5.53080E+03    2.10278E-03 +/-  0.497 %
    4    6.02360E-01    9.15021E-04 +/-  0.534 %
    5    1.00000E-05    3.68919E-04 +/-  0.825 %

 << VOIDED REGION EDGE CELL SPCTRUM JUST INSIDE VOIDED REGION ON AM >>
 GROUP    ELOWER(EV)        FLUX
    1    1.00000E+07    2.01849E-05 +/-  9.655 %
    2    8.20850E+05    3.07985E-03 +/-  0.369 %
    3    5.53080E+03    2.10340E-03 +/-  0.314 %
    4    6.02360E-01    9.27934E-04 +/-  0.347 %
    5    1.00000E-05    3.69902E-04 +/-  0.442 %

 << UNVOIDED REGION EDGE CELL SPCTRUM JUST OUTSIDE VOIDED REGION ON AM >>
 GROUP    ELOWER(EV)        FLUX
    1    1.00000E+07    2.05074E-05 +/-  8.475 %
    2    8.20850E+05    3.08867E-03 +/-  0.363 %
    3    5.53080E+03    2.11875E-03 +/-  0.288 %
    4    6.02360E-01    9.32265E-04 +/-  0.309 %
    5    1.00000E-05    3.69285E-04 +/-  0.488 %

 << VOIDED REGION EDGE CELL SPCTRUM JUST INSIDE VOIDED REGION ON AD >>
 GROUP    ELOWER(EV)        FLUX
    1    1.00000E+07    1.96311E-05 +/- 11.178 %
    2    8.20850E+05    3.06883E-03 +/-  0.423 %
    3    5.53080E+03    2.10463E-03 +/-  0.350 %
    4    6.02360E-01    9.24019E-04 +/-  0.419 %
    5    1.00000E-05    3.70520E-04 +/-  0.642 %

 << UNVOIDED REGION EDGE CELL SPCTRUM JUST OUTSIDE VOIDED REGION ON AD >>
 GROUP    ELOWER(EV)        FLUX
    1    1.00000E+07    2.03919E-05 +/- 10.011 %
    2    8.20850E+05    3.07412E-03 +/-  0.351 %
    3    5.53080E+03    2.09960E-03 +/-  0.422 %
    4    6.02360E-01    9.32321E-04 +/-  0.396 %
    5    1.00000E-05    3.72173E-04 +/-  0.673 %

 << UNVOIDED CELL SPCTRUM (AT D) >>
 GROUP    ELOWER(EV)        FLUX
    1    1.00000E+07    1.87302E-05 +/- 14.399 %
    2    8.20850E+05    3.07131E-03 +/-  0.671 %
    3    5.53080E+03    2.10907E-03 +/-  0.556 %
    4    6.02360E-01    9.20484E-04 +/-  0.550 %
    5    1.00000E-05    3.69732E-04 +/-  0.946 %

## NEACRP PWR VOID COEFFICIENT BENCHMARK BY MVP ## UO2 VOIDED

 FLUX DENSITY , PER SOURCE NETUTRON , PER LETHARGY

 << VOIDED REGION CENTRAL CELL SPECTRUM >>
 GROUP    ELOWER(EV)        FLUX
    1    1.00000E+07    2.12381E-05 +/- 13.313 %
    2    8.20850E+05    3.27282E-03 +/-  0.549 %
    3    5.53080E+03    2.87318E-03 +/-  0.425 %
    4    6.02360E-01    7.83396E-04 +/-  0.588 %
    5    1.00000E-05    1.08345E-04 +/-  1.341 %

 << VOIDED REGION EDGE CELL SPCTRUM JUST INSIDE VOIDED REGION ON AM >>
 GROUP    ELOWER(EV)        FLUX
    1    1.00000E+07    1.88508E-05 +/- 10.867 %
    2    8.20850E+05    3.22150E-03 +/-  0.358 %
    3    5.53080E+03    2.68678E-03 +/-  0.345 %
    4    6.02360E-01    8.68603E-04 +/-  0.344 %
    5    1.00000E-05    1.62334E-04 +/-  0.830 %

 << UNVOIDED REGION EDGE CELL SPCTRUM JUST OUTSIDE VOIDED REGION ON AM >>
 GROUP    ELOWER(EV)        FLUX
    1    1.00000E+07    1.73711E-05 +/- 10.695 %
    2    8.20850E+05    3.11589E-03 +/-  0.361 %
    3    5.53080E+03    2.55259E-03 +/-  0.310 %
    4    6.02360E-01    9.42547E-04 +/-  0.339 %
    5    1.00000E-05    2.45139E-04 +/-  0.687 %

 << VOIDED REGION EDGE CELL SPCTRUM JUST INSIDE VOIDED REGION ON AD >>
 GROUP    ELOWER(EV)        FLUX
    1    1.00000E+07    1.75938E-05 +/- 10.828 %
    2    8.20850E+05    3.22909E-03 +/-  0.487 %
    3    5.53080E+03    2.58220E-03 +/-  0.355 %
    4    6.02360E-01    9.19230E-04 +/-  0.378 %
    5    1.00000E-05    2.14124E-04 +/-  0.775 %

 << UNVOIDED REGION EDGE CELL SPCTRUM JUST OUTSIDE VOIDED REGION ON AD >>
 GROUP    ELOWER(EV)        FLUX
    1    1.00000E+07    2.02002E-05 +/- 11.259 %
    2    8.20850E+05    3.18894E-03 +/-  0.383 %
    3    5.53080E+03    2.41343E-03 +/-  0.373 %
    4    6.02360E-01    9.96030E-04 +/-  0.394 %
    5    1.00000E-05    3.29470E-04 +/-  0.767 %

 << UNVOIDED CELL SPCTRUM (AT D) >>
 GROUP    ELOWER(EV)        FLUX
    1    1.00000E+07    1.84120E-05 +/- 14.268 %
    2    8.20850E+05    3.41033E-03 +/-  0.553 %
    3    5.53080E+03    2.32110E-03 +/-  0.508 %
    4    6.02360E-01    1.02103E-03 +/-  0.610 %
    5    1.00000E-05    4.09710E-04 +/-  0.931 %

## NEACRP PWR VOID COEFFICIENT BENCHMARK BY MVP ## H-MOX UNVOIDED

 FLUX DENSITY , PER SOURCE NETUTRON , PER LETHARGY

 << VOIDED REGION CENTRAL CELL SPECTRUM >>
 GROUP    ELOWER(EV)        FLUX
    1    1.00000E+07    3.18868E-05 +/- 14.126 %
    2    8.20850E+05    3.10451E-03 +/-  0.579 %
    3    5.53080E+03    2.03344E-03 +/-  0.526 %
    4    6.02360E-01    7.70731E-04 +/-  0.657 %
    5    1.00000E-05    5.40595E-05 +/-  1.826 %

 << VOIDED REGION EDGE CELL SPCTRUM JUST INSIDE VOIDED REGION ON AM >>
 GROUP    ELOWER(EV)        FLUX
    1    1.00000E+07    2.57041E-05 +/-  8.479 %
    2    8.20850E+05    3.21906E-03 +/-  0.384 %
    3    5.53080E+03    2.09221E-03 +/-  0.308 %
    4    6.02360E-01    8.18300E-04 +/-  0.360 %
    5    1.00000E-05    1.04686E-04 +/-  0.744 %

 << UNVOIDED REGION EDGE CELL SPCTRUM JUST OUTSIDE VOIDED REGION ON AM >>
 GROUP    ELOWER(EV)        FLUX
    1    1.00000E+07    2.55472E-05 +/-  7.335 %
    2    8.20850E+05    3.01697E-03 +/-  0.370 %
    3    5.53080E+03    2.05898E-03 +/-  0.309 %
    4    6.02360E-01    8.71841E-04 +/-  0.348 %
    5    1.00000E-05    2.28382E-04 +/-  0.709 %
```

```
<< VOIDED REGION EDGE CELL SPCTRUM JUST INSIDE VOIDED REGION ON AD >>
 GROUP    ELOWER(EV)       FLUX
    1   1.00000E+07    3.58058E-05 +/-  9.854 %
    2   8.20850E+05    3.29797E-03 +/-  0.449 %
    3   5.53080E+03    2.12009E-03 +/-  0.390 %
    4   6.02360E-01    8.31504E-04 +/-  0.448 %
    5   1.00000E-05    1.48848E-04 +/-  0.715 %

<< UNVOIDED REGION EDGE CELL SPCTRUM JUST OUTSIDE VOIDED REGION ON AD >>
 GROUP    ELOWER(EV)       FLUX
    1   1.00000E+07    2.59006E-05 +/-  8.929 %
    2   8.20850E+05    3.02673E-03 +/-  0.430 %
    3   5.53080E+03    2.06663E-03 +/-  0.375 %
    4   6.02360E-01    8.93583E-04 +/-  0.386 %
    5   1.00000E-05    3.03133E-04 +/-  0.757 %

<< UNVOIDED CELL SPCTRUM (AT D) >>
 GROUP    ELOWER(EV)       FLUX
    1   1.00000E+07    1.89760E-05 +/- 15.496 %
    2   8.20850E+05    3.11868E-03 +/-  0.592 %
    3   5.53080E+03    2.12610E-03 +/-  0.496 %
    4   6.02360E-01    9.41310E-04 +/-  0.521 %
    5   1.00000E-05    3.73476E-04 +/-  0.759 %


## NEACRP PWR VOID COEFFICIENT BENCHMARK BY MVP ## H-MOX VOIDED

 FLUX DENSITY , PER SOURCE NETUTRON , PER LETHARGY

<< VOIDED REGION CENTRAL CELL SPECTRUM >>
 GROUP    ELOWER(EV)       FLUX
    1   1.00000E+07    2.30886E-05 +/- 14.084 %
    2   8.20850E+05    3.47679E-03 +/-  0.559 %
    3   5.53080E+03    2.87125E-03 +/-  0.499 %
    4   6.02360E-01    5.90740E-04 +/-  0.652 %
    5   1.00000E-05    6.89307E-06 +/-  5.318 %

<< VOIDED REGION EDGE CELL SPCTRUM JUST INSIDE VOIDED REGION ON AM >>
 GROUP    ELOWER(EV)       FLUX
    1   1.00000E+07    2.23882E-05 +/-  8.672 %
    2   8.20850E+05    3.50905E-03 +/-  0.344 %
    3   5.53080E+03    2.72477E-03 +/-  0.307 %
    4   6.02360E-01    7.13760E-04 +/-  0.393 %
    5   1.00000E-05    5.09901E-05 +/-  1.110 %

<< UNVOIDED REGION EDGE CELL SPCTRUM JUST OUTSIDE VOIDED REGION ON AM >>
 GROUP    ELOWER(EV)       FLUX
    1   1.00000E+07    2.30901E-05 +/-  8.816 %
    2   8.20850E+05    3.21565E-03 +/-  0.365 %
    3   5.53080E+03    2.56195E-03 +/-  0.290 %
    4   6.02360E-01    8.52539E-04 +/-  0.346 %
    5   1.00000E-05    1.65901E-04 +/-  0.701 %

<< VOIDED REGION EDGE CELL SPCTRUM JUST INSIDE VOIDED REGION ON AD >>
 GROUP    ELOWER(EV)       FLUX
    1   1.00000E+07    2.34194E-05 +/-  9.447 %
    2   8.20850E+05    3.52368E-03 +/-  0.422 %
    3   5.53080E+03    2.63416E-03 +/-  0.366 %
    4   6.02360E-01    8.05679E-04 +/-  0.459 %
    5   1.00000E-05    9.83188E-05 +/-  0.853 %

<< UNVOIDED REGION EDGE CELL SPCTRUM JUST OUTSIDE VOIDED REGION ON AD >>
 GROUP    ELOWER(EV)       FLUX
    1   1.00000E+07    2.07545E-05 +/-  9.725 %
    2   8.20850E+05    3.24128E-03 +/-  0.404 %
    3   5.53080E+03    2.41686E-03 +/-  0.387 %
    4   6.02360E-01    9.60340E-04 +/-  0.401 %
    5   1.00000E-05    2.80847E-04 +/-  0.706 %
```

```
<< UNVOIDED CELL SPCTRUM (AT D) >>
 GROUP    ELOWER(EV)       FLUX
    1   1.00000E+07    2.18817E-05 +/- 15.154 %
    2   8.20850E+05    3.40309E-03 +/-  0.563 %
    3   5.53080E+03    2.32291E-03 +/-  0.513 %
    4   6.02360E-01    1.02192E-03 +/-  0.590 %
    5   1.00000E-05    4.09561E-04 +/-  0.949 %

## NEACRP PWR VOID COEFFICIENT BENCHMARK BY MVP ## M-MOX UNVOIDED

 FLUX DENSITY , PER SOURCE NETUTRON , PER LETHARGY

<< VOIDED REGION CENTRAL CELL SPECTRUM >>
 GROUP    ELOWER(EV)       FLUX
    1   1.00000E+07    2.14973E-05 +/- 13.443 %
    2   8.20850E+05    3.02841E-03 +/-  0.547 %
    3   5.53080E+03    2.01184E-03 +/-  0.495 %
    4   6.02360E-01    8.00707E-04 +/-  0.615 %
    5   1.00000E-05    7.44300E-05 +/-  1.455 %

<< VOIDED REGION EDGE CELL SPCTRUM JUST INSIDE VOIDED REGION ON AM >>
 GROUP    ELOWER(EV)       FLUX
    1   1.00000E+07    2.93853E-05 +/-  7.630 %
    2   8.20850E+05    3.13426E-03 +/-  0.349 %
    3   5.53080E+03    2.04525E-03 +/-  0.287 %
    4   6.02360E-01    8.38954E-04 +/-  0.347 %
    5   1.00000E-05    1.29318E-04 +/-  0.746 %

<< UNVOIDED REGION EDGE CELL SPCTRUM JUST OUTSIDE VOIDED REGION ON AM >>
 GROUP    ELOWER(EV)       FLUX
    1   1.00000E+07    2.81303E-05 +/-  9.649 %
    2   8.20850E+05    2.99853E-03 +/-  0.336 %
    3   5.53080E+03    2.03303E-03 +/-  0.339 %
    4   6.02360E-01    8.78366E-04 +/-  0.374 %
    5   1.00000E-05    2.41596E-04 +/-  0.686 %

<< VOIDED REGION EDGE CELL SPCTRUM JUST INSIDE VOIDED REGION ON AD >>
 GROUP    ELOWER(EV)       FLUX
    1   1.00000E+07    2.57227E-05 +/-  9.968 %
    2   8.20850E+05    3.23155E-03 +/-  0.453 %
    3   5.53080E+03    2.08043E-03 +/-  0.382 %
    4   6.02360E-01    8.56325E-04 +/-  0.438 %
    5   1.00000E-05    1.71955E-04 +/-  0.791 %

<< UNVOIDED REGION EDGE CELL SPCTRUM JUST OUTSIDE VOIDED REGION ON AD >>
 GROUP    ELOWER(EV)       FLUX
    1   1.00000E+07    2.25282E-05 +/-  9.184 %
    2   8.20850E+05    3.02736E-03 +/-  0.458 %
    3   5.53080E+03    2.08278E-03 +/-  0.382 %
    4   6.02360E-01    9.02879E-04 +/-  0.394 %
    5   1.00000E-05    3.08472E-04 +/-  0.647 %

<< UNVOIDED CELL SPCTRUM (AT D) >>
 GROUP    ELOWER(EV)       FLUX
    1   1.00000E+07    2.47058E-05 +/- 14.966 %
    2   8.20850E+05    3.16588E-03 +/-  0.597 %
    3   5.53080E+03    2.15387E-03 +/-  0.497 %
    4   6.02360E-01    9.39823E-04 +/-  0.568 %
    5   1.00000E-05    3.72626E-04 +/-  0.799 %


## NEACRP PWR VOID COEFFICIENT BENCHMARK BY MVP ## M-MOX VOIDED

 FLUX DENSITY , PER SOURCE NETUTRON , PER LETHARGY

<< VOIDED REGION CENTRAL CELL SPECTRUM >>
```

```
 GROUP    ELOWER(EV)      FLUX
    1    1.00000E+07   2.19448E-05 +/- 13.386 %
    2    8.20850E+05   3.34041E-03 +/-  0.587 %
    3    5.53080E+03   2.83202E-03 +/-  0.464 %
    4    6.02360E-01   6.37491E-04 +/-  0.658 %
    5    1.00000E-05   1.11240E-05 +/-  3.830 %

<< VOIDED REGION EDGE CELL SPCTRUM JUST INSIDE VOIDED REGION ON AM >>
 GROUP    ELOWER(EV)      FLUX
    1    1.00000E+07   2.25019E-05 +/-  8.929 %
    2    8.20850E+05   3.39260E-03 +/-  0.378 %
    3    5.53080E+03   2.68743E-03 +/-  0.291 %
    4    6.02360E-01   7.54472E-04 +/-  0.391 %
    5    1.00000E-05   6.17443E-05 +/-  0.902 %

<< UNVOIDED REGION EDGE CELL SPCTRUM JUST OUTSIDE VOIDED REGION ON AM >>
 GROUP    ELOWER(EV)      FLUX
    1    1.00000E+07   1.95167E-05 +/-  8.736 %
    2    8.20850E+05   3.17759E-03 +/-  0.416 %
    3    5.53080E+03   2.54397E-03 +/-  0.299 %
    4    6.02360E-01   8.80199E-04 +/-  0.363 %
    5    1.00000E-05   1.74615E-04 +/-  0.670 %

<< VOIDED REGION EDGE CELL SPCTRUM JUST INSIDE VOIDED REGION ON AD >>
 GROUP    ELOWER(EV)      FLUX
    1    1.00000E+07   2.57124E-05 +/-  8.924 %
    2    8.20850E+05   3.44638E-03 +/-  0.440 %
    3    5.53080E+03   2.59244E-03 +/-  0.336 %
    4    6.02360E-01   8.35516E-04 +/-  0.461 %
    5    1.00000E-05   1.12887E-04 +/-  0.873 %

<< UNVOIDED REGION EDGE CELL SPCTRUM JUST OUTSIDE VOIDED REGION ON AD >>
 GROUP    ELOWER(EV)      FLUX
    1    1.00000E+07   2.66928E-05 +/-  8.758 %
    2    8.20850E+05   3.21982E-03 +/-  0.438 %
    3    5.53080E+03   2.41947E-03 +/-  0.374 %
    4    6.02360E-01   9.73113E-04 +/-  0.427 %
    5    1.00000E-05   2.86348E-04 +/-  0.668 %

<< UNVOIDED CELL SPCTRUM (AT D) >>
 GROUP    ELOWER(EV)      FLUX
    1    1.00000E+07   2.73598E-05 +/- 11.784 %
    2    8.20850E+05   3.41852E-03 +/-  0.500 %
    3    5.53080E+03   2.33888E-03 +/-  0.531 %
    4    6.02360E-01   1.02431E-03 +/-  0.516 %
    5    1.00000E-05   4.01404E-04 +/-  0.833 %

## NEACRP PWR VOID COEFFICIENT BENCHMARK BY MVP ## L-MOX UNVOIDED

  FLUX DENSITY , PER SOURCE NETUTRON , PER LETHARGY

<< VOIDED REGION CENTRAL CELL SPECTRUM >>
 GROUP    ELOWER(EV)      FLUX
    1    1.00000E+07   2.31056E-05 +/- 15.243 %
    2    8.20850E+05   2.96804E-03 +/-  0.644 %
    3    5.53080E+03   1.98507E-03 +/-  0.482 %
    4    6.02360E-01   8.43056E-04 +/-  0.542 %
    5    1.00000E-05   1.18906E-04 +/-  1.334 %

<< VOIDED REGION EDGE CELL SPCTRUM JUST INSIDE VOIDED REGION ON AM >>
 GROUP    ELOWER(EV)      FLUX
    1    1.00000E+07   2.70547E-05 +/-  7.713 %
    2    8.20850E+05   3.06023E-03 +/-  0.406 %
    3    5.53080E+03   2.03474E-03 +/-  0.292 %
    4    6.02360E-01   8.61754E-04 +/-  0.315 %
    5    1.00000E-05   1.75110E-04 +/-  0.678 %
```

```
<< UNVOIDED REGION EDGE CELL SPCTRUM JUST OUTSIDE VOIDED REGION ON AM >>
 GROUP    ELOWER(EV)      FLUX
    1    1.00000E+07   2.33165E-05 +/-  8.240 %
    2    8.20850E+05   2.96817E-03 +/-  0.327 %
    3    5.53080E+03   2.03652E-03 +/-  0.318 %
    4    6.02360E-01   8.90601E-04 +/-  0.302 %
    5    1.00000E-05   2.69119E-04 +/-  0.624 %

<< VOIDED REGION EDGE CELL SPCTRUM JUST INSIDE VOIDED REGION ON AD >>
 GROUP    ELOWER(EV)      FLUX
    1    1.00000E+07   2.23130E-05 +/- 10.064 %
    2    8.20850E+05   3.12447E-03 +/-  0.404 %
    3    5.53080E+03   2.06489E-03 +/-  0.364 %
    4    6.02360E-01   8.79998E-04 +/-  0.408 %
    5    1.00000E-05   2.15684E-04 +/-  0.614 %

<< UNVOIDED REGION EDGE CELL SPCTRUM JUST OUTSIDE VOIDED REGION ON AD >>
 GROUP    ELOWER(EV)      FLUX
    1    1.00000E+07   2.17723E-05 +/- 10.752 %
    2    8.20850E+05   3.01803E-03 +/-  0.402 %
    3    5.53080E+03   2.05436E-03 +/-  0.407 %
    4    6.02360E-01   9.08908E-04 +/-  0.377 %
    5    1.00000E-05   3.23363E-04 +/-  0.588 %

<< UNVOIDED CELL SPCTRUM (AT D) >>
 GROUP    ELOWER(EV)      FLUX
    1    1.00000E+07   2.03254E-05 +/- 13.430 %
    2    8.20850E+05   3.20155E-03 +/-  0.553 %
    3    5.53080E+03   2.14144E-03 +/-  0.521 %
    4    6.02360E-01   9.41833E-04 +/-  0.555 %
    5    1.00000E-05   3.79011E-04 +/-  0.805 %

## NEACRP PWR VOID COEFFICIENT BENCHMARK BY MVP ## L-MOX VOIDED

  FLUX DENSITY , PER SOURCE NETUTRON , PER LETHARGY

<< VOIDED REGION CENTRAL CELL SPECTRUM >>
 GROUP    ELOWER(EV)      FLUX
    1    1.00000E+07   1.92796E-05 +/- 14.871 %
    2    8.20850E+05   3.24191E-03 +/-  0.537 %
    3    5.53080E+03   2.81816E-03 +/-  0.450 %
    4    6.02360E-01   7.08777E-04 +/-  0.736 %
    5    1.00000E-05   2.51309E-05 +/-  2.539 %

<< VOIDED REGION EDGE CELL SPCTRUM JUST INSIDE VOIDED REGION ON AM >>
 GROUP    ELOWER(EV)      FLUX
    1    1.00000E+07   2.04950E-05 +/- 10.352 %
    2    8.20850E+05   3.27348E-03 +/-  0.354 %
    3    5.53080E+03   2.68182E-03 +/-  0.324 %
    4    6.02360E-01   8.00879E-04 +/-  0.374 %
    5    1.00000E-05   7.90514E-05 +/-  0.990 %

<< UNVOIDED REGION EDGE CELL SPCTRUM JUST OUTSIDE VOIDED REGION ON AM >>
 GROUP    ELOWER(EV)      FLUX
    1    1.00000E+07   2.19677E-05 +/-  8.914 %
    2    8.20850E+05   3.09905E-03 +/-  0.351 %
    3    5.53080E+03   2.53268E-03 +/-  0.309 %
    4    6.02360E-01   9.00573E-04 +/-  0.380 %
    5    1.00000E-05   1.86973E-04 +/-  0.704 %

<< VOIDED REGION EDGE CELL SPCTRUM JUST INSIDE VOIDED REGION ON AD >>
 GROUP    ELOWER(EV)      FLUX
    1    1.00000E+07   2.01107E-05 +/- 10.385 %
    2    8.20850E+05   3.32955E-03 +/-  0.383 %
    3    5.53080E+03   2.58700E-03 +/-  0.371 %
    4    6.02360E-01   8.68770E-04 +/-  0.432 %
    5    1.00000E-05   1.38106E-04 +/-  0.956 %
```

```
<< UNVOIDED REGION EDGE CELL SPCTRUM JUST OUTSIDE VOIDED REGION ON AD >>
 GROUP    ELOWER(EV)        FLUX
    1    1.00000E+07    1.91956E-05 +/- 11.079 %
    2    8.20850E+05    3.16386E-03 +/-  0.405 %
    3    5.53080E+03    2.40035E-03 +/-  0.367 %
    4    6.02360E-01    9.90750E-04 +/-  0.398 %
    5    1.00000E-05    3.01670E-04 +/-  0.714 %

<< UNVOIDED CELL SPCTRUM (AT D) >>
 GROUP    ELOWER(EV)        FLUX
    1    1.00000E+07    2.47019E-05 +/- 12.707 %
    2    8.20850E+05    3.41741E-03 +/-  0.545 %
    3    5.53080E+03    2.34776E-03 +/-  0.469 %
    4    6.02360E-01    1.01758E-03 +/-  0.551 %
    5    1.00000E-05    4.11166E-04 +/-  0.845 %
```

Appendix B.9

Results of void reactivity effect benchmark

K. Okumura (JAERI)

From JAERI, two sets of results have been already submitted by H. Takano et al. (see Appendix B.8). There is a result based on a continuous-energy Monte Carlo method by the MVP code and a result based on a combinational method with a collision probability calculation for the unit cell and a finite difference diffusion calculation for the macrocell by SRAC code.

Here two sets of results based on other methods for the macrocell calculation are presented. One is based on a collision probability method (SRAC-PIJ) and the other is based on a modern nodal diffusion method (MOSRA).

SRAC-PIJ

- *Calculational method, program name and version*: SRAC [1] is a code system which has the functions of a collision probability (PIJ), Sn calculations (ANISN, TWOTRAN) and finite difference diffusion calculations (TUD, CITATION). A direct collision probability calculation for the heterogeneous macrocell on a fine energy group structure is too time-consuming. Therefore, the macrocell calculation was carried out by PIJ with the homogenised macroscopic cross-sections (107 groups), which were produced by PIJ in the unit cell. The outer boundary condition of the macrocell was approximated by isotropic (white) reflection to reduce computational time.

- *Data libraries and original evaluated data file*: SRACLIB-JENDL3 processed JENDL-3.1 data file.

- *List of isotopes for resonance shielding calculation and method used*: the resonance treatment in the unit cell calculation is the same in the case of SRAC submitted by H. Takano et al. (see Appendix B.8).

MOSRA

- *Calculational method, program name and version*: MOSRA [2] is a nuclear design code system for various types of advanced reactors. It is still under development but some neutronics modules are available. For this benchmark study, a nodal diffusion calculation module for a vector processor was employed. It is based on a fourth-order polynomial nodal expansion method (NEM) with a quadratic transverse leakage approximation [3]. The macroscopic cross-sections homogenised in the unit cell were produced by collision probability calculations of the SRAC code. The macrocell calculations were carried out by MOSRA on 107 groups with the mesh size equal to the cell pitch (1.26cm).

- *Data libraries and original evaluated data file*: SRACLIB-JENDL3 processed from JENDL-3.1 data file.

- *List of isotopes for resonance shielding calculation and method used*: the resonance treatment in the unit cell calculation is the same in the case of SRAC submitted by H. Takano et al.

References

[1] K. Tsuchihashi et al., "Revised SRAC Code System", JAERI-1302 (1986).

[2] K. Okumura, "Development of Nuclear Design Code System for Advanced Reactors", published in "Reactor Engineering Department Annual Report (April 1st 1993 – March 31st 1994)", JAERI Review (1994).

[3] F. Finnemann, F. Bennewitz and M. R. Wagner, "Interface Current Techniques for Multi-dimensional Reactor Calculations", Atomkernenergie, 30 123-128, (1977).

```
## NEACRP PWR VOID COEFFICIENT BENCHMARK BY SRAC-Pij ## H-MOX UNVOIDED

   RESULT BY SRAC-DIFF.: UPPER ENERGY LIMIT  IS 10000000. EV
                       : No. OF ENERGY GROUPS IS 107G(LATTICE)/107G(MACROCELL)
                       : NO AXIAL BUCKLING

 H-MOX UNVOIDED
   FLUX DENSITY , PER SOURCE NETUTRON , PER LETHARGY

 << VOIDED REGION CENTRAL CELL SPECTRUM >>
  GROUP    ELOWER(EV)      FLUX
     1   8.20850E+05   3.16615E-03
     2   5.53084E+03   2.06535E-03
     3   6.02360E-01   7.80565E-04
     4   1.00000E-05   5.68710E-05

 << VOIDED REGION EDGE CELL SPCTRUM JUST INSIDE VOIDED REGION ON AM >>
  GROUP    ELOWER(EV)      FLUX
     1   8.20850E+05   3.26167E-03
     2   5.53084E+03   2.09311E-03
     3   6.02360E-01   8.16884E-04
     4   1.00000E-05   1.10960E-04

 << UNVOIDED REGION EDGE CELL SPCTRUM JUST OUTSIDE VOIDED REGION ON AM >>
  GROUP    ELOWER(EV)      FLUX
     1   8.20850E+05   3.07609E-03
     2   5.53084E+03   2.06165E-03
     3   6.02360E-01   8.67601E-04
     4   1.00000E-05   2.20441E-04

 << VOIDED REGION EDGE CELL SPCTRUM JUST INSIDE VOIDED REGION ON AD >>
  GROUP    ELOWER(EV)      FLUX
     1   8.20850E+05   3.31526E-03
     2   5.53084E+03   2.10971E-03
     3   6.02360E-01   8.42500E-04
     4   1.00000E-05   1.57717E-04

 << UNVOIDED REGION EDGE CELL SPCTRUM JUST OUTSIDE VOIDED REGION ON AD >>
  GROUP    ELOWER(EV)      FLUX
     1   8.20850E+05   3.07483E-03
     2   5.53084E+03   2.06931E-03
     3   6.02360E-01   8.97584E-04
     4   1.00000E-05   2.98052E-04

 << UNVOIDED CELL SPCTRUM (AT D) >>
  GROUP    ELOWER(EV)      FLUX
     1   8.20850E+05   3.15552E-03
     2   5.53084E+03   2.12049E-03
     3   6.02360E-01   9.32319E-04
     4   1.00000E-05   3.69997E-04
H-MOX VOIDED

   FLUX DENSITY , PER SOURCE NETUTRON , PER LETHARGY

 << VOIDED REGION CENTRAL CELL SPECTRUM >>
  GROUP    ELOWER(EV)      FLUX
     1   8.20850E+05   3.52892E-03
     2   5.53084E+03   2.89246E-03
     3   6.02360E-01   5.87974E-04
     4   1.00000E-05   1.63697E-06

 << VOIDED REGION EDGE CELL SPCTRUM JUST INSIDE VOIDED REGION ON AM >>
  GROUP    ELOWER(EV)      FLUX
     1   8.20850E+05   3.58915E-03
     2   5.53084E+03   2.74140E-03
     3   6.02360E-01   7.17237E-04
     4   1.00000E-05   3.68997E-05

 << UNVOIDED REGION EDGE CELL SPCTRUM JUST OUTSIDE VOIDED REGION ON AM >>
  GROUP    ELOWER(EV)      FLUX
     1   8.20850E+05   3.28644E-03
     2   5.53084E+03   2.55634E-03
     3   6.02360E-01   8.65562E-04
     4   1.00000E-05   1.66070E-04

 << VOIDED REGION EDGE CELL SPCTRUM JUST INSIDE VOIDED REGION ON AD >>
  GROUP    ELOWER(EV)      FLUX
     1   8.20850E+05   3.66268E-03
     2   5.53084E+03   2.63531E-03
     3   6.02360E-01   8.05201E-04
     4   1.00000E-05   7.70403E-05

 << UNVOIDED REGION EDGE CELL SPCTRUM JUST OUTSIDE VOIDED REGION ON AD >>
  GROUP    ELOWER(EV)      FLUX
     1   8.20850E+05   3.27830E-03
     2   5.53084E+03   2.41240E-03
     3   6.02360E-01   9.60719E-04
     4   1.00000E-05   2.79583E-04

 << UNVOIDED CELL SPCTRUM (AT D) >>
  GROUP    ELOWER(EV)      FLUX
     1   8.20850E+05   3.41998E-03
     2   5.53084E+03   2.30786E-03
     3   6.02360E-01   1.01664E-03
     4   1.00000E-05   4.04154E-04
L-MOX UNVOIDED

   FLUX DENSITY , PER SOURCE NETUTRON , PER LETHARGY

 << VOIDED REGION CENTRAL CELL SPECTRUM >>
  GROUP    ELOWER(EV)      FLUX
     1   8.20850E+05   2.99680E-03
     2   5.53084E+03   2.00038E-03
     3   6.02360E-01   8.46247E-04
     4   1.00000E-05   1.23225E-04

 << VOIDED REGION EDGE CELL SPCTRUM JUST INSIDE VOIDED REGION ON AM >>
  GROUP    ELOWER(EV)      FLUX
     1   8.20850E+05   3.09641E-03
     2   5.53084E+03   2.04076E-03
     3   6.02360E-01   8.65893E-04
     4   1.00000E-05   1.80527E-04

 << UNVOIDED REGION EDGE CELL SPCTRUM JUST OUTSIDE VOIDED REGION ON AM >>
  GROUP    ELOWER(EV)      FLUX
     1   8.20850E+05   3.01895E-03
     2   5.53084E+03   2.03490E-03
     3   6.02360E-01   8.90134E-04
     4   1.00000E-05   2.63238E-04

 << VOIDED REGION EDGE CELL SPCTRUM JUST INSIDE VOIDED REGION ON AD >>
  GROUP    ELOWER(EV)      FLUX
     1   8.20850E+05   3.15752E-03
     2   5.53084E+03   2.06858E-03
     3   6.02360E-01   8.80658E-04
     4   1.00000E-05   2.24628E-04

 << UNVOIDED REGION EDGE CELL SPCTRUM JUST OUTSIDE VOIDED REGION ON AD >>
  GROUP    ELOWER(EV)      FLUX
     1   8.20850E+05   3.06559E-03
     2   5.53084E+03   2.06424E-03
```

```
      3    6.02360E-01    9.08240E-04
      4    1.00000E-05    3.21655E-04

  << UNVOIDED CELL SPCTRUM (AT D) >>
   GROUP    ELOWER(EV)       FLUX
      1    8.20850E+05    3.18848E-03
      2    5.53084E+03    2.14277E-03
      3    6.02360E-01    9.41513E-04
      4    1.00000E-05    3.73528E-04

L-MOX VOIDED

   FLUX DENSITY , PER SOURCE NETUTRON , PER LETHARGY

  << VOIDED REGION CENTRAL CELL SPECTRUM >>
   GROUP    ELOWER(EV)       FLUX
      1    8.20850E+05    3.28895E-03
      2    5.53084E+03    2.84289E-03
      3    6.02360E-01    7.03295E-04
      4    1.00000E-05    1.54714E-05

  << VOIDED REGION EDGE CELL SPCTRUM JUST INSIDE VOIDED REGION ON AM >>
   GROUP    ELOWER(EV)       FLUX
      1    8.20850E+05    3.34630E-03
      2    5.53084E+03    2.68447E-03
      3    6.02360E-01    8.02792E-04
      4    1.00000E-05    6.88339E-05

  << UNVOIDED REGION EDGE CELL SPCTRUM JUST OUTSIDE VOIDED REGION ON AM >>
   GROUP    ELOWER(EV)       FLUX
      1    8.20850E+05    3.15739E-03
      2    5.53084E+03    2.51198E-03
      3    6.02360E-01    9.13030E-04
      4    1.00000E-05    1.87112E-04

  << VOIDED REGION EDGE CELL SPCTRUM JUST INSIDE VOIDED REGION ON AD >>
   GROUP    ELOWER(EV)       FLUX
      1    8.20850E+05    3.41718E-03
      2    5.53084E+03    2.57871E-03
      3    6.02360E-01    8.69457E-04
      4    1.00000E-05    1.23977E-04

  << UNVOIDED REGION EDGE CELL SPCTRUM JUST OUTSIDE VOIDED REGION ON AD >>
   GROUP    ELOWER(EV)       FLUX
      1    8.20850E+05    3.22189E-03
      2    5.53084E+03    2.38934E-03
      3    6.02360E-01    9.81924E-04
      4    1.00000E-05    2.96379E-04

  << UNVOIDED CELL SPCTRUM (AT D) >>
   GROUP    ELOWER(EV)       FLUX
      1    8.20850E+05    3.45815E-03
      2    5.53084E+03    2.33290E-03
      3    6.02360E-01    1.02648E-03
      4    1.00000E-05    4.07841E-04

M-MOX UNVOIDED

   FLUX DENSITY , PER SOURCE NETUTRON , PER LETHARGY

  << VOIDED REGION CENTRAL CELL SPECTRUM >>
   GROUP    ELOWER(EV)       FLUX
      1    8.20850E+05    3.07273E-03
      2    5.53084E+03    2.02958E-03
      3    6.02360E-01    8.09564E-04
      4    1.00000E-05    7.70074E-05

  << VOIDED REGION EDGE CELL SPCTRUM JUST INSIDE VOIDED REGION ON AM >>
   GROUP    ELOWER(EV)       FLUX
      1    8.20850E+05    3.17926E-03
      2    5.53084E+03    2.06600E-03
      3    6.02360E-01    8.38652E-04
      4    1.00000E-05    1.33937E-04

  << UNVOIDED REGION EDGE CELL SPCTRUM JUST OUTSIDE VOIDED REGION ON AM >>
   GROUP    ELOWER(EV)       FLUX
      1    8.20850E+05    3.04191E-03
      2    5.53084E+03    2.04608E-03
      3    6.02360E-01    8.77425E-04
      4    1.00000E-05    2.35410E-04

  << VOIDED REGION EDGE CELL SPCTRUM JUST INSIDE VOIDED REGION ON AD >>
   GROUP    ELOWER(EV)       FLUX
      1    8.20850E+05    3.24175E-03
      2    5.53084E+03    2.08958E-03
      3    6.02360E-01    8.59602E-04
      4    1.00000E-05    1.81068E-04

  << UNVOIDED REGION EDGE CELL SPCTRUM JUST OUTSIDE VOIDED REGION ON AD >>
   GROUP    ELOWER(EV)       FLUX
      1    8.20850E+05    3.06709E-03
      2    5.53084E+03    2.06553E-03
      3    6.02360E-01    9.02300E-04
      4    1.00000E-05    3.06881E-04

  << UNVOIDED CELL SPCTRUM (AT D) >>
   GROUP    ELOWER(EV)       FLUX
      1    8.20850E+05    3.17577E-03
      2    5.53084E+03    2.13394E-03
      3    6.02360E-01    9.37762E-04
      4    1.00000E-05    3.72055E-04

M- VOIDED

   FLUX DENSITY , PER SOURCE NETUTRON , PER LETHARGY

  << VOIDED REGION CENTRAL CELL SPECTRUM >>
   GROUP    ELOWER(EV)       FLUX
      1    8.20850E+05    3.40099E-03
      2    5.53084E+03    2.86668E-03
      3    6.02360E-01    6.39961E-04
      4    1.00000E-05    4.62278E-06

  << VOIDED REGION EDGE CELL SPCTRUM JUST INSIDE VOIDED REGION ON AM >>
   GROUP    ELOWER(EV)       FLUX
      1    8.20850E+05    3.47033E-03
      2    5.53084E+03    2.71380E-03
      3    6.02360E-01    7.55951E-04
      4    1.00000E-05    4.82021E-05

  << UNVOIDED REGION EDGE CELL SPCTRUM JUST OUTSIDE VOIDED REGION ON AM >>
   GROUP    ELOWER(EV)       FLUX
      1    8.20850E+05    3.22083E-03
      2    5.53084E+03    2.53410E-03
      3    6.02360E-01    8.87254E-04
      4    1.00000E-05    1.73672E-04

  << VOIDED REGION EDGE CELL SPCTRUM JUST INSIDE VOIDED REGION ON AD >>
   GROUP    ELOWER(EV)       FLUX
      1    8.20850E+05    3.54846E-03
      2    5.53084E+03    2.60895E-03
      3    6.02360E-01    8.34393E-04
      4    1.00000E-05    9.55581E-05

  << UNVOIDED REGION EDGE CELL SPCTRUM JUST OUTSIDE VOIDED REGION ON AD >>
   GROUP    ELOWER(EV)       FLUX
```

```
      1    8.20850E+05    3.24981E-03
      2    5.53084E+03    2.40098E-03
      3    6.02360E-01    9.70507E-04
      4    1.00000E-05    2.86271E-04

  << UNVOIDED CELL SPCTRUM (AT D) >>
   GROUP    ELOWER(EV)       FLUX
      1    8.20850E+05    3.44084E-03
      2    5.53084E+03    2.32146E-03
      3    6.02360E-01    1.02195E-03
      4    1.00000E-05    4.06145E-04

UO2 UNVOIDED

   FLUX DENSITY , PER SOURCE NETUTRON , PER LETHARGY

  << VOIDED REGION CENTRAL CELL SPECTRUM >>
   GROUP    ELOWER(EV)       FLUX
      1    8.20850E+05    3.11412E-03
      2    5.53084E+03    2.09855E-03
      3    6.02360E-01    9.25411E-04
      4    1.00000E-05    3.68527E-04

  << VOIDED REGION EDGE CELL SPCTRUM JUST INSIDE VOIDED REGION ON AM >>
   GROUP    ELOWER(EV)       FLUX
      1    8.20850E+05    3.11401E-03
      2    5.53084E+03    2.09814E-03
      3    6.02360E-01    9.25094E-04
      4    1.00000E-05    3.68321E-04

  << UNVOIDED REGION EDGE CELL SPCTRUM JUST OUTSIDE VOIDED REGION ON AM >>
   GROUP    ELOWER(EV)       FLUX
      1    8.20850E+05    3.11396E-03
      2    5.53084E+03    2.09816E-03
      3    6.02360E-01    9.25106E-04
      4    1.00000E-05    3.68330E-04

  << VOIDED REGION EDGE CELL SPCTRUM JUST INSIDE VOIDED REGION ON AD >>
   GROUP    ELOWER(EV)       FLUX
      1    8.20850E+05    3.11409E-03
      2    5.53084E+03    2.09811E-03
      3    6.02360E-01    9.24938E-04
      4    1.00000E-05    3.68232E-04

  << UNVOIDED REGION EDGE CELL SPCTRUM JUST OUTSIDE VOIDED REGION ON AD >>
   GROUP    ELOWER(EV)       FLUX
      1    8.20850E+05    3.11396E-03
      2    5.53084E+03    2.09819E-03
      3    6.02360E-01    9.24985E-04
      4    1.00000E-05    3.68262E-04

  << UNVOIDED CELL SPCTRUM (AT D) >>
   GROUP    ELOWER(EV)       FLUX
      1    8.20850E+05    3.11408E-03
      2    5.53084E+03    2.09842E-03
      3    6.02360E-01    9.25323E-04
      4    1.00000E-05    3.68517E-04

UO2 VOIDED

   FLUX DENSITY , PER SOURCE NETUTRON , PER LETHARGY

  << VOIDED REGION CENTRAL CELL SPECTRUM >>
   GROUP    ELOWER(EV)       FLUX
      1    8.20850E+05    3.31921E-03
      2    5.53084E+03    2.88434E-03
      3    6.02360E-01    7.88007E-04
      4    1.00000E-05    9.83167E-05

  << VOIDED REGION EDGE CELL SPCTRUM JUST INSIDE VOIDED REGION ON AM >>
   GROUP    ELOWER(EV)       FLUX
      1    8.20850E+05    3.29045E-03
      2    5.53084E+03    2.70330E-03
      3    6.02360E-01    8.67334E-04
      4    1.00000E-05    1.53549E-04

  << UNVOIDED REGION EDGE CELL SPCTRUM JUST OUTSIDE VOIDED REGION ON AM >>
   GROUP    ELOWER(EV)       FLUX
      1    8.20850E+05    3.16666E-03
      2    5.53084E+03    2.53669E-03
      3    6.02360E-01    9.46413E-04
      4    1.00000E-05    2.43473E-04

  << VOIDED REGION EDGE CELL SPCTRUM JUST INSIDE VOIDED REGION ON AD >>
   GROUP    ELOWER(EV)       FLUX
      1    8.20850E+05    3.30804E-03
      2    5.53084E+03    2.58171E-03
      3    6.02360E-01    9.19609E-04
      4    1.00000E-05    2.06546E-04

  << UNVOIDED REGION EDGE CELL SPCTRUM JUST OUTSIDE VOIDED REGION ON AD >>
   GROUP    ELOWER(EV)       FLUX
      1    8.20850E+05    3.22968E-03
      2    5.53084E+03    2.40466E-03
      3    6.02360E-01    9.97657E-04
      4    1.00000E-05    3.27991E-04

  << UNVOIDED CELL SPCTRUM (AT D) >>
   GROUP    ELOWER(EV)       FLUX
      1    8.20850E+05    3.43064E-03
      2    5.53084E+03    2.31771E-03
      3    6.02360E-01    1.02138E-03
      4    1.00000E-05    4.06433E-04
```

```
NEACRP PWR : H-MOX UNVOIDED CORE :  1.26 CM PITCH  : 2D-XY QUATAR CORE

  RESULT BY MOSRA       : UPPER ENERGY LIMIT  IS 10000000. EV
     (HOMOGENEOUS)      : NO OF ENERGY GROUPS IS 107G(LATTICE)/107G(MACROCELL)

 FLUX DENSITY , PER SOURCE NETUTRON , PER LETHARGY

<< VOIDED REGION CENTRAL CELL SPECTRUM >>
 GROUP   ELOWER(EV)      FLUX
    1   8.20850E+05   3.14489E-03
    2   5.53084E+03   2.05257E-03
    3   6.02360E-01   7.74457E-04
    4   1.00000E-05   5.55670E-05

<< VOIDED REGION EDGE CELL SPCTRUM JUST INSIDE VOIDED REGION ON AM >>
 GROUP   ELOWER(EV)      FLUX
    1   8.20850E+05   3.15508E-03
    2   5.53084E+03   2.06870E-03
    3   6.02360E-01   8.17642E-04
    4   1.00000E-05   1.11818E-04

<< UNVOIDED REGION EDGE CELL SPCTRUM JUST OUTSIDE VOIDED REGION ON AM >>
 GROUP   ELOWER(EV)      FLUX
    1   8.20850E+05   3.10995E-03
    2   5.53084E+03   2.06198E-03
    3   6.02360E-01   8.59993E-04
    4   1.00000E-05   2.13306E-04

<< VOIDED REGION EDGE CELL SPCTRUM JUST INSIDE VOIDED REGION ON AD >>
 GROUP   ELOWER(EV)      FLUX
    1   8.20850E+05   3.15849E-03
    2   5.53084E+03   2.07971E-03
    3   6.02360E-01   8.47883E-04
    4   1.00000E-05   1.62067E-04

<< UNVOIDED REGION EDGE CELL SPCTRUM JUST OUTSIDE VOIDED REGION ON AD >>
 GROUP   ELOWER(EV)      FLUX
    1   8.20850E+05   3.10274E-03
    2   5.53084E+03   2.07220E-03
    3   6.02360E-01   8.94001E-04
    4   1.00000E-05   2.92541E-04

<< UNVOIDED CELL SPCTRUM (AT D) >>
 GROUP   ELOWER(EV)      FLUX
    1   8.20850E+05   3.15803E-03
    2   5.53084E+03   2.12324E-03
    3   6.02360E-01   9.33662E-04
    4   1.00000E-05   3.70981E-04
NEACRP PWR : H-MOX VOIDED CORE :  1.26 CM PITCH  : 2D-XY QUATAR CORE

 FLUX DENSITY , PER SOURCE NETUTRON , PER LETHARGY

<< VOIDED REGION CENTRAL CELL SPECTRUM >>
 GROUP   ELOWER(EV)      FLUX
    1   8.20850E+05   3.45787E-03
    2   5.53084E+03   2.76010E-03
    3   6.02360E-01   6.18265E-04
    4   1.00000E-05   1.00515E-06

<< VOIDED REGION EDGE CELL SPCTRUM JUST INSIDE VOIDED REGION ON AM >>
 GROUP   ELOWER(EV)      FLUX
    1   8.20850E+05   3.41748E-03
    2   5.53084E+03   2.64481E-03
    3   6.02360E-01   7.40215E-04
    4   1.00000E-05   3.78034E-05

<< UNVOIDED REGION EDGE CELL SPCTRUM JUST OUTSIDE VOIDED REGION ON AM >>
 GROUP   ELOWER(EV)      FLUX
    1   8.20850E+05   3.34069E-03
    2   5.53084E+03   2.55291E-03
    3   6.02360E-01   8.49573E-04
    4   1.00000E-05   1.50654E-04

<< VOIDED REGION EDGE CELL SPCTRUM JUST INSIDE VOIDED REGION ON AD >>
 GROUP   ELOWER(EV)      FLUX
    1   8.20850E+05   3.40077E-03
    2   5.53084E+03   2.54966E-03
    3   6.02360E-01   8.28864E-04
    4   1.00000E-05   8.19305E-05

<< UNVOIDED REGION EDGE CELL SPCTRUM JUST OUTSIDE VOIDED REGION ON AD >>
 GROUP   ELOWER(EV)      FLUX
    1   8.20850E+05   3.31253E-03
    2   5.53084E+03   2.41807E-03
    3   6.02360E-01   9.56223E-04
    4   1.00000E-05   2.68358E-04

<< UNVOIDED CELL SPCTRUM (AT D) >>
 GROUP   ELOWER(EV)      FLUX
    1   8.20850E+05   3.41629E-03
    2   5.53084E+03   2.30468E-03
    3   6.02360E-01   1.01527E-03
    4   1.00000E-05   4.04224E-04


NEACRP PWR : L-MOX UNVOIDED CORE :  1.26 CM PITCH  : 2D-XY QUATAR CORE


 FLUX DENSITY , PER SOURCE NETUTRON , PER LETHARGY

<< VOIDED REGION CENTRAL CELL SPECTRUM >>
 GROUP   ELOWER(EV)      FLUX
    1   8.20850E+05   3.01558E-03
    2   5.53084E+03   2.00616E-03
    3   6.02360E-01   8.46350E-04
    4   1.00000E-05   1.21304E-04

<< VOIDED REGION EDGE CELL SPCTRUM JUST INSIDE VOIDED REGION ON AM >>
 GROUP   ELOWER(EV)      FLUX
    1   8.20850E+05   3.05953E-03
    2   5.53084E+03   2.03602E-03
    3   6.02360E-01   8.67983E-04
    4   1.00000E-05   1.81535E-04

<< UNVOIDED REGION EDGE CELL SPCTRUM JUST OUTSIDE VOIDED REGION ON AM >>
 GROUP   ELOWER(EV)      FLUX
    1   8.20850E+05   3.05149E-03
    2   5.53084E+03   2.04103E-03
    3   6.02360E-01   8.88387E-04
    4   1.00000E-05   2.58074E-04

<< VOIDED REGION EDGE CELL SPCTRUM JUST INSIDE VOIDED REGION ON AD >>
 GROUP   ELOWER(EV)      FLUX
    1   8.20850E+05   3.08947E-03
    2   5.53084E+03   2.05816E-03
    3   6.02360E-01   8.84098E-04
    4   1.00000E-05   2.28614E-04

<< UNVOIDED REGION EDGE CELL SPCTRUM JUST OUTSIDE VOIDED REGION ON AD >>
 GROUP   ELOWER(EV)      FLUX
    1   8.20850E+05   3.08366E-03
    2   5.53084E+03   2.06739E-03
```

```
     3    6.02360E-01    9.07447E-04
     4    1.00000E-05    3.18489E-04

 << UNVOIDED CELL SPCTRUM (AT D) >>
  GROUP    ELOWER(EV)       FLUX
     1    8.20850E+05    3.18623E-03
     2    5.53084E+03    2.14245E-03
     3    6.02360E-01    9.41539E-04
     4    1.00000E-05    3.73955E-04
```

NEACRP PWR : L-MOX VOIDED CORE : 1.26 CM PITCH : 2D-XY QUATAR CORE

FLUX DENSITY , PER SOURCE NETUTRON , PER LETHARGY

```
 << VOIDED REGION CENTRAL CELL SPECTRUM >>
  GROUP    ELOWER(EV)       FLUX
     1    8.20850E+05    3.28924E-03
     2    5.53084E+03    2.72552E-03
     3    6.02360E-01    7.34283E-04
     4    1.00000E-05    1.29675E-05

 << VOIDED REGION EDGE CELL SPCTRUM JUST INSIDE VOIDED REGION ON AM >>
  GROUP    ELOWER(EV)       FLUX
     1    8.20850E+05    3.27862E-03
     2    5.53084E+03    2.60935E-03
     3    6.02360E-01    8.24273E-04
     4    1.00000E-05    6.94764E-05

 << UNVOIDED REGION EDGE CELL SPCTRUM JUST OUTSIDE VOIDED REGION ON AM >>
  GROUP    ELOWER(EV)       FLUX
     1    8.20850E+05    3.23636E-03
     2    5.53084E+03    2.52131E-03
     3    6.02360E-01    9.03826E-04
     4    1.00000E-05    1.71678E-04

 << VOIDED REGION EDGE CELL SPCTRUM JUST INSIDE VOIDED REGION ON AD >>
  GROUP    ELOWER(EV)       FLUX
     1    8.20850E+05    3.28902E-03
     2    5.53084E+03    2.51778E-03
     3    6.02360E-01    8.89171E-04
     4    1.00000E-05    1.27702E-04

 << UNVOIDED REGION EDGE CELL SPCTRUM JUST OUTSIDE VOIDED REGION ON AD >>
  GROUP    ELOWER(EV)       FLUX
     1    8.20850E+05    3.26171E-03
     2    5.53084E+03    2.40065E-03
     3    6.02360E-01    9.80816E-04
     4    1.00000E-05    2.88476E-04

 << UNVOIDED CELL SPCTRUM (AT D) >>
  GROUP    ELOWER(EV)       FLUX
     1    8.20850E+05    3.45113E-03
     2    5.53084E+03    2.32780E-03
     3    6.02360E-01    1.02431E-03
     4    1.00000E-05    4.07482E-04
```

NEACRP PWR : M-MOX UNVOIDED CORE : 1.26 CM PITCH : 2D-XY QUATAR CORE

FLUX DENSITY , PER SOURCE NETUTRON , PER LETHARGY

```
 << VOIDED REGION CENTRAL CELL SPECTRUM >>
  GROUP    ELOWER(EV)       FLUX
     1    8.20850E+05    3.07434E-03
     2    5.53084E+03    2.02667E-03
     3    6.02360E-01    8.06463E-04
     4    1.00000E-05    7.55329E-05

 << VOIDED REGION EDGE CELL SPCTRUM JUST INSIDE VOIDED REGION ON AM >>
  GROUP    ELOWER(EV)       FLUX
     1    8.20850E+05    3.10468E-03
     2    5.53084E+03    2.05084E-03
     3    6.02360E-01    8.40120E-04
     4    1.00000E-05    1.34941E-04

 << UNVOIDED REGION EDGE CELL SPCTRUM JUST OUTSIDE VOIDED REGION ON AM >>
  GROUP    ELOWER(EV)       FLUX
     1    8.20850E+05    3.07728E-03
     2    5.53084E+03    2.04973E-03
     3    6.02360E-01    8.72579E-04
     4    1.00000E-05    2.28980E-04

 << VOIDED REGION EDGE CELL SPCTRUM JUST INSIDE VOIDED REGION ON AD >>
  GROUP    ELOWER(EV)       FLUX
     1    8.20850E+05    3.12328E-03
     2    5.53084E+03    2.06818E-03
     3    6.02360E-01    8.64152E-04
     4    1.00000E-05    1.85434E-04

 << UNVOIDED REGION EDGE CELL SPCTRUM JUST OUTSIDE VOIDED REGION ON AD >>
  GROUP    ELOWER(EV)       FLUX
     1    8.20850E+05    3.09132E-03
     2    5.53084E+03    2.06879E-03
     3    6.02360E-01    9.00047E-04
     4    1.00000E-05    3.02315E-04

 << UNVOIDED CELL SPCTRUM (AT D) >>
  GROUP    ELOWER(EV)       FLUX
     1    8.20850E+05    3.17552E-03
     2    5.53084E+03    2.13497E-03
     3    6.02360E-01    9.38381E-04
     4    1.00000E-05    3.72734E-04
```

NEACRP PWR : M-MOX VOIDED CORE : 1.26 CM PITCH : 2D-XY QUATAR CORE

FLUX DENSITY , PER SOURCE NETUTRON , PER LETHARGY

```
 << VOIDED REGION CENTRAL CELL SPECTRUM >>
  GROUP    ELOWER(EV)       FLUX
     1    8.20850E+05    3.37010E-03
     2    5.53084E+03    2.74214E-03
     3    6.02360E-01    6.71077E-04
     4    1.00000E-05    3.29196E-06

 << VOIDED REGION EDGE CELL SPCTRUM JUST INSIDE VOIDED REGION ON AM >>
  GROUP    ELOWER(EV)       FLUX
     1    8.20850E+05    3.34752E-03
     2    5.53084E+03    2.62711E-03
     3    6.02360E-01    7.78488E-04
     4    1.00000E-05    4.92158E-05

 << UNVOIDED REGION EDGE CELL SPCTRUM JUST OUTSIDE VOIDED REGION ON AM >>
  GROUP    ELOWER(EV)       FLUX
     1    8.20850E+05    3.28754E-03
     2    5.53084E+03    2.53691E-03
     3    6.02360E-01    8.74418E-04
     4    1.00000E-05    1.58093E-04

 << VOIDED REGION EDGE CELL SPCTRUM JUST INSIDE VOIDED REGION ON AD >>
  GROUP    ELOWER(EV)       FLUX
     1    8.20850E+05    3.34611E-03
     2    5.53084E+03    2.53421E-03
     3    6.02360E-01    8.56361E-04
```

```
    4   1.00000E-05    1.00546E-04

 << UNVOIDED REGION EDGE CELL SPCTRUM JUST OUTSIDE VOIDED REGION ON AD >>
  GROUP    ELOWER(EV)       FLUX
     1    8.20850E+05    3.28687E-03
     2    5.53084E+03    2.40933E-03
     3    6.02360E-01    9.67620E-04
     4    1.00000E-05    2.76451E-04

 << UNVOIDED CELL SPCTRUM (AT D) >>
  GROUP    ELOWER(EV)       FLUX
     1    8.20850E+05    3.43556E-03
     2    5.53084E+03    2.31739E-03
     3    6.02360E-01    1.02021E-03
     4    1.00000E-05    4.06001E-04

NEACRP PWR : UO2 UNVOIDED CORE :   1.26 CM PITCH   : 2D-XY QUATAR CORE


  FLUX DENSITY , PER SOURCE NETUTRON , PER LETHARGY

 << VOIDED REGION CENTRAL CELL SPECTRUM >>
  GROUP    ELOWER(EV)       FLUX
     1    8.20850E+05    3.11372E-03
     2    5.53084E+03    2.09829E-03
     3    6.02360E-01    9.25319E-04
     4    1.00000E-05    3.68474E-04

 << VOIDED REGION EDGE CELL SPCTRUM JUST INSIDE VOIDED REGION ON AM >>
  GROUP    ELOWER(EV)       FLUX
     1    8.20850E+05    3.11373E-03
     2    5.53084E+03    2.09828E-03
     3    6.02360E-01    9.25320E-04
     4    1.00000E-05    3.68475E-04

 << UNVOIDED REGION EDGE CELL SPCTRUM JUST OUTSIDE VOIDED REGION ON AM >>
  GROUP    ELOWER(EV)       FLUX
     1    8.20850E+05    3.11371E-03
     2    5.53084E+03    2.09829E-03
     3    6.02360E-01    9.25319E-04
     4    1.00000E-05    3.68475E-04

 << VOIDED REGION EDGE CELL SPCTRUM JUST INSIDE VOIDED REGION ON AD >>
  GROUP    ELOWER(EV)       FLUX
     1    8.20850E+05    3.11373E-03
     2    5.53084E+03    2.09829E-03
     3    6.02360E-01    9.25320E-04
     4    1.00000E-05    3.68475E-04

 << UNVOIDED REGION EDGE CELL SPCTRUM JUST OUTSIDE VOIDED REGION ON AD >>
  GROUP    ELOWER(EV)       FLUX
     1    8.20850E+05    3.11372E-03
     2    5.53084E+03    2.09829E-03
     3    6.02360E-01    9.25320E-04
     4    1.00000E-05    3.68475E-04

 << UNVOIDED CELL SPCTRUM (AT D) >>
  GROUP    ELOWER(EV)       FLUX
     1    8.20850E+05    3.11373E-03
     2    5.53084E+03    2.09829E-03
     3    6.02360E-01    9.25321E-04
     4    1.00000E-05    3.68476E-04

NEACRP PWR : UO2 VOIDED CORE :   1.26 CM PITCH   : 2D-XY QUATAR CORE

  FLUX DENSITY , PER SOURCE NETUTRON , PER LETHARGY

 << VOIDED REGION CENTRAL CELL SPECTRUM >>
  GROUP    ELOWER(EV)       FLUX
     1    8.20850E+05    3.30845E-03
     2    5.53084E+03    2.76458E-03
     3    6.02360E-01    8.18933E-04
     4    1.00000E-05    1.00573E-04

 << VOIDED REGION EDGE CELL SPCTRUM JUST INSIDE VOIDED REGION ON AM >>
  GROUP    ELOWER(EV)       FLUX
     1    8.20850E+05    3.27761E-03
     2    5.53084E+03    2.63723E-03
     3    6.02360E-01    8.86705E-04
     4    1.00000E-05    1.54526E-04

 << UNVOIDED REGION EDGE CELL SPCTRUM JUST OUTSIDE VOIDED REGION ON AM >>
  GROUP    ELOWER(EV)       FLUX
     1    8.20850E+05    3.24618E-03
     2    5.53084E+03    2.54847E-03
     3    6.02360E-01    9.41306E-04
     4    1.00000E-05    2.32868E-04

 << VOIDED REGION EDGE CELL SPCTRUM JUST INSIDE VOIDED REGION ON AD >>
  GROUP    ELOWER(EV)       FLUX
     1    8.20850E+05    3.27523E-03
     2    5.53084E+03    2.53632E-03
     3    6.02360E-01    9.35519E-04
     4    1.00000E-05    2.05457E-04

 << UNVOIDED REGION EDGE CELL SPCTRUM JUST OUTSIDE VOIDED REGION ON AD >>
  GROUP    ELOWER(EV)       FLUX
     1    8.20850E+05    3.26777E-03
     2    5.53084E+03    2.41797E-03
     3    6.02360E-01    9.98459E-04
     4    1.00000E-05    3.24708E-04

 << UNVOIDED CELL SPCTRUM (AT D) >>
  GROUP    ELOWER(EV)       FLUX
     1    8.20850E+05    3.42273E-03
     2    5.53084E+03    2.31143E-03
     3    6.02360E-01    1.01870E-03
     4    1.00000E-05    4.05780E-04
```

Appendix B.10

Void reactivity effect benchmark

Y. Uenohara (Toshiba Corporation)

Calculational method	**Continuous-energy Monte Carlo**
Name of code	**MCNP**
Version No	**4.2**
Data libraries	the authors have produced ACE-formatted libraries with the process code **NJOY** version 91.13. See JAERI-M 93-046, 169-177, for detailed information.
Evaluated data file	**JENDL version 3.1**
Resonance treatment	Self-shielding effects derived from "Resolved Resonances" have been considered fully. However, those from "Unresolved Resonances" were ignored. In analyses of usual LWRs, the self-shielding effects from unresolved resonances scarcely contribute to the results. Therefore infinite diluted cross-sections were used in this case.

Appendix C.1

Physical analysis of void reactivity effect benchmark problem

S. Cathalau, M. Soldevila, S. Rahlfs, A. Maghnouj, G. Rimpault, P. J. Finck (CEA)

General analysis

We first carried out a comparison between all solutions provided by participants, results for the infinite lattice are given in Table C.1-1 below.

CASE	UO_2	L-MOX	M-MOX	H-MOX
$k_{unvoided}$	1.3656 ± 0.0054	1.1506 ± 0.0099	1.1772 ± 0.0087	1.2146 ± 0.0074
k_{voided}	0.6278 ± 0.0090	0.7618 ± 0.0088	1.0324 ± 0.0100	1.2748 ± 0.0107
void effect *	- 0.8608 ± 0.0206	- 0.4437 ± 0.0203	- 0.1156 ± 0.0135	0.0385 ± 0.0092

* obtained by averaging the reactivity effects ($1/k_{unvoided} - 1/k_{voided}$)

Table C.1-1 ***Average k-infinity for infinite lattice calculations***

Results spread up to 1% (1σ) for the multiplication factor, and up to 2 % (1σ) for the void effect. They show the same tendency as in benchmark problem 1A and 1B.

Restricting the analysis to calculations performed with the same evaluated data file (JEF-2.2), the spreads are approximately divided by 2 as shown in Table C.1-2 below.

CASE	UO_2	L-MOX	M-MOX	H-MOX
$k_{unvoided}$	1.3732 ± 0.0020	1.1482 ± 0.0025	1.1703 ± 0.0022	1.2116 ± 0.0039
k_{voided}	0.6441 ± 0.0049	0.7733 ± 0.0029	1.0448 ± 0.0020	1.2864 ± 0.0012
void effect *	- 0.8242 ± 0.0119	- 0.4222 ± 0.0056	- 0.1027 ± 0.0025	0.0467 ± 0.0016

Table C.1-2 ***Average k-infinity for infinite lattice calculations (JEF-2.2)***

The reactivity effect for infinite lattice is strongly negative for the uranium lattice (due to a large negative contribution of U-238); it is also negative for the "classical" L-MOX (about 5% plutonium content) lattice and the M-MOX fuel (9% Pu) in which the positive contributions of Pu-240 and Pu-242 (loss of the capture reactions in the first resonances in the thermal energy range) cannot compensate for the negative contribution of U-238. For the H-MOX lattice (14% Pu), these positive contributions are higher than the negative contributions, resulting in a positive void reactivity effect. For the macrocell, the average results including all solutions are the following:

CASE	UO_2	L-MOX	M-MOX	H-MOX
$k_{unvoided}$	1.3660 ± 0.0061	1.3387 ± 0.0072	1.3395 ± 0.0071	1.3429 ± 0.0071
k_{voided}	1.3513 ± 0.0061	1.3392 ± 0.0082	1.3427 ± 0.0080	1.3472 ± 0.0078
void effect	- 0.00797 ± 0.00059	0.00029 ± 0.00133	0.00178 ± 0.00124	0.00235 ± 0.00108

Table C.1-3 ***Average k-infinity for macrocell calculations***

The spread is less than 0.8% (1σ) for the multiplication factors and less than 0.2% for the void effect. Restricting the analysis to calculations performed with the same evaluated data file (JEF-2.2), the spreads are strongly reduced as shown in Table C.1-4 below.

CASE	UO_2	L-MOX	M-MOX	H-MOX
$k_{unvoided}$	1.3735 ± 0.0028	1.3452 ± 0.0024	1.3460 ± 0.0026	1.3437 ± 0.0020
k_{voided}	1.3591 ± 0.0023	1.3469 ± 0.0027	1.3506 ± 0.0031	1.3549 ± 0.0027
void effect	- 0.0077 ± 0.0003	0.0009 ± 0.0003	0.0025 ± 0.0006	0.0030 ± 0.0005

Table C.1-4 ***Average k-infinity for macrocell calculations (JEF-2.2)***

The unexpected results are the positive reactivity void effects for the M-MOX macrocell and particularly for the L-MOX macrocell.

Physical analysis

Several analyses were performed using standard calculations and Perturbation Theory (*PT*). The first analysis was performed for the variation of k* ($\nu\ \Sigma_f / \Sigma_a$). The second one deals with a qualitative analysis of the direct and adjoint fluxes for the unvoided and voided macrocells. Finally, a perturbation calculation was performed to provide energy and isotopic contributions to the reactivity effects.

Neutronic balance using k*

Calculations were performed by the APOLLO-2 code using the CEA-93 library (172 groups), UP1 approximation for the collision probability calculations and 900 physical cells (in which the flux is calculated). Six spatial regions were used in each fuel pin.

Infinite lattice calculations give the following k-infinity:

Flooded UO_2 cell	$k^u{}_f = 1.3746$
Flooded L-MOX cell	$k^m{}_f = 1.1496$
Voided L-MOX cell	$k^m{}_v = 0.7766$

For the L-MOX macrocell, we got the following results:

Flooded	$k_f = 1.34644$
Voided	$k_v = 1.34761$

So, a positive void coefficient of $\Delta\rho = 1/k_f - 1/k_v = +64$ pcm is obtained.

Table C.1-5 and Figure C.1-1 give the variation of the $k^*(r) = \nu\ \Sigma_f(r) / \Sigma_a(r)$ along the median axis AM of the macrocell for the voided and the flooded cases. The dotted line is relative to the flooded case and the continuous line stands for the voided case.

CELL	FLOODED CASE	VOIDED CASE
1	1.1559	1.1143
2	1.1597	1.1255
3	1.1700	1.1485
4	1.1927	1.1834
5	1.2337	1.2060
6	1.2914	1.2107
7	1.3306	1.2800
8	1.3504	1.3189
9	1.3608	1.3409
10	1.3663	1.3539
11	1.3692	1.3616
12	1.3707	1.3662
13	1.3715	1.3690
14	1.3718	1.3705
15	1.3719	1.3711

Table C.1-5 ***Variation of k* along the median axis L-MOX macrocell***

We can note that k* for both flooded and voided cases in the UO_2 zone is decreasing along the median axis when approaching the MOX zone; for the corner cell, we observe almost the same value as for the infinite lattice calculation (k-infinity = 1.37); $k^*_{UO_2}$ decreases to 1.29 for the flooded case and to 1.21 for the voided situation.

For the MOX zone, we observe that k^*_{MOX} is very similar in both voided and flooded cases and the average value is very close to the infinite lattice value which is about 1.15. This indicates that, in the flooded case, the central cell is almost in the fundamental code whereas it is very far from this situation in the voided case; in the latter case, the spectrum within the MOX zone is driven by thermal neutrons streaming from the border cells of the UO_2 zone.

From these calculations, we can also try to establish a neutronic balance in the voided and flooded cases. Let us introduce the following notations:

k^*_f	average $\nu\Sigma_f/\Sigma_a$	for the *flooded* macrocell
k^*_v	average $\nu\Sigma_f/\Sigma_a$	for the *voided* macrocell
k^u_f	average $\nu\Sigma_f/\Sigma_a$	for the *UO_2* zone in the *flooded* macrocell
k^u_v	average $\nu\Sigma_f/\Sigma_a$	for the *UO_2* zone in the *voided* macrocell

k^m_f	average $\nu\Sigma_f/\Sigma_a$	for the *MOX* zone in the *flooded* macrocell
k^m_v	average $\nu\Sigma_f/\Sigma_a$	for the *MOX* zone in the *voided* macrocell

And the neutronic weight of each zone are:

ω^u_f	contribution of the UO_2 zone in the *flooded* case
ω^u_v	contribution of the UO_2 zone in the *voided* case
ω^m_f	contribution of the *MOX* zone in the *flooded* case
ω^m_f	contribution of the *MOX* zone in the *flooded* case

with $\omega^u_f + \omega^m_f = 1.0$ and $\omega^u_v + \omega^m_v = 1.0$

and
$$k^*_f = \omega^u_f . k^u_f + \omega^m_f . k^m_f$$
$$k^*_v = \omega^u_v . k^u_v + \omega^m_v . k^m_v$$

When solving this linear system, we obtain the following contributions in the voided and flooded situations:

CASE	UO_2 ZONE	MOX ZONE
flooded	0.871	0.129
voided	0.931	0.069

Table C.1-6

Table C.1-6 shows that in the voided case the contribution of the MOX to the averaged k* is two times lower than in the flooded situation, and consequently the infinite multiplication factor of the voided macrocell gets closer to the one of the infinite UO_2 lattice.

Figure C.1-2 shows the variations of the k* factors of the macrocells and of each zone in the voided and flooded situations. k* of both UO_2 and MOX zones are decreasing when the moderator is removed, the variation of the contributions of each zone result (by linear combination) in an increase of the k* of the macrocell.

In order to understand this phenomenon, we can introduce the following notations:

$k^u_v = k^u_f + \delta^u$
$k^m_v = k^m_f + \delta^m$ where δ^u, δ^m and δ^ω are negative.
$\omega^u_v = \omega^u_f + \delta^\omega$
$\omega^m_v = \omega^m_f + \delta^\omega$

The void reactivity effect can be obtained using the following formula:

$$\delta\rho = \frac{1}{k_v k_f} \times \left[-\delta^{\omega} \times \left(k^u{}_f - k^m{}_f \right) + \omega^u{}_f \times \delta^u + \omega^m{}_f \times \delta^m - \delta^{\omega} \times \left(\delta^u - \delta^m \right) \right]$$

A *B* *C* *D*

where	*A*	stands for	the difference between the flooded UO_2 and MOX average k*
	B	is	the weighted variation of the UO_2 average k*
	C	is	the weighted variation of the MOX average k*
	D	is	a second order effect.

Numerically we obtain:

Flooded case $k^1{}_f = 1.3648$
$k^2{}_f = 1.2081$

Voided case $k^1{}_v = 1.3564$
$k^2{}_v = 1.2001$

Consequently $\delta u = -0.0080$
$\delta m = -0.0074$
$\delta\omega = -0.060$

Thus, the neutronic balance can be written:

A	=	+0.00940
B	=	-0.00740
C	=	-0.00103
D	=	+0.00007

Total = +0.00099 ⇒ $\delta\rho = 55$ pcm

The main positive contribution to the void coefficient comes from the difference between the flooded UO_2 and MOX average k* factors; this contribution is balanced by the negative contributions of the decrease of the UO_2 and MOX average k* factor when the moderator of the central zone is removed.

Figures C.1-3 and C.1-4 show the absorption rates in the macrocell with the total production rate normalised to 1.0. We can see an important increase of the absorption in the UO_2 region when the moderator is removed, which causes an increase of the production rate in the UO_2 zone,

Figures C.1-5 and C.1-6 present the variations of the capture and fission rates in the macrocell when we remove the moderator (the productions rates are also normalised to 1.0). We can observe a very large decrease of the capture and fission rates in the MOX zone and consequently an increase of these reaction rates in the UO_2 cells.

This study shows that the void reactivity effect comes from the rebalancing of the reaction rates (absorption and consequently production rates) from the central MOX zone to the external UO_2 zone.

The calculation of the neutronic weights for the flooded and the voided situation shows a decrease by a factor of about 2 of the "importance" of the central MOX zone when the moderator is removed.

Qualitative analyses of the adjoint fluxes in the L-MOX macrocell

This analysis consists in characterising the variation of the fluxes in each zone with regard to the corresponding adjoint fluxes, when the moderator is removed in the central L-MOX zone.

We chose to analyse the spectra for the following cells:

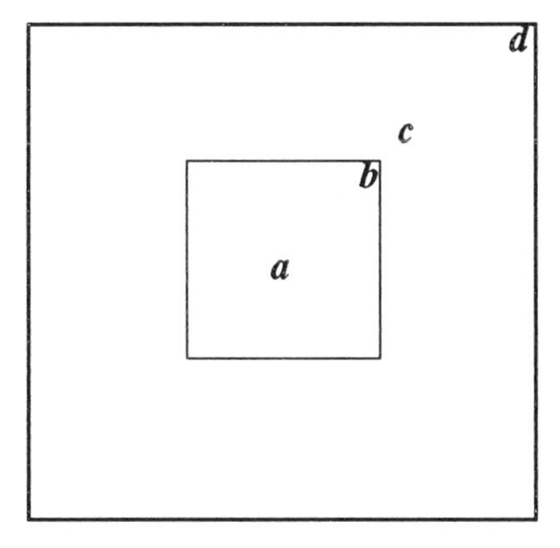

where
- *a* central L-MOX cell,
- *b* L-MOX cell placed at the corner of the MOX zone and against the interface with the UO_2 zone,
- *c* UO_2 cell symmetric to b on the diagonal axis,
- *d* "asymptotic" UO_2 cell at the corner on the diagonal axis.

Figure C.1-7 shows the variation of the direct flux along the diagonal axis (the integrated fluxes are normalised to 1.0) in the flooded situation, and in particular the modification of the level of the thermal flux between cell "d" and the central MOX cell "a"; the Maxwellian shape has almost disappeared.

Figure C.1-8 shows the fluxes in the voided configuration. Several phenomena are occurring:

- The level of the thermal flux in cells "c" and "d" is somewhat higher than in the flooded case,

- The thermal flux has vanished in the central MOX cell "a",
- A very large increase of the fast flux for the central cell,
- A strong modification of the slope in the slowing-down energy range for the flux in the central cell, due to the loss of the moderator in the central zone.

However, as shown in Figure C.1-9, the flux in the central MOX zone of the macrocell (for cells "a" and "b") do not correspond to the fundamental spectrum of a voided MOX cell placed in an infinite medium: the neutrons coming from the UO_2 zone are driving the MOX spectrum.

Figures C.1-10 and C.1-11 show the adjoint fluxes in the flooded and voided situation respectively. We can note the following phenomena:

- The large resonances of Pu-240, Pu-242 and U-238 are clearly shown on those curves: this indicates that, if a neutron arises at the corresponding energy, its capture probability is very large;
- In the flooded case, we can note that the importance of the "asymptotic" UO_2 cell is higher than the others all over the energy range;
- For the voided situation we note that, for the central L-MOX cell, the importance in the low thermal range (under the Maxwellian peak) is higher than the others. It could be explained by the fact that, if a neutron arises at this energy, it will be directly absorbed by the plutonium isotopes (thus producing new neutrons). We do not observe this phenomena in the flooded situation because the moderator causes up-scattering (Maxwellian effect) resulting in a lower probability for this neutron to be absorbed by the plutonium isotopes.

Figures C.1-12 and C.1-13 show the total importance for the two situations, calculated as $\Phi(u),\Phi^*(u)$. We can observe the following phenomena;

- In the thermal range, the total importances of the UO_2 cells are the same;
- A decrease by a factor of 2 of the cell "b" importance (L-MOX at the comer of the L-MOX zone) in the thermal range;
- A very large decrease of the importance of the central L-MOX cells and the thermal range;
- Consequently, the importance in the fast energy range increases for the cell "a".

In conclusion, this qualitative analysis allows to improve our understanding of the distribution of the neutrons in the macrocell when the moderator is removed. We have shown that, in the flooded case, the thermal spectrum for the L-MOX cells are lower than the one obtained for the UO_2 cells consequently, the fast spectrum is higher.

In the voided case, the most important phenomena are linked to the "rebalancing" of the reaction rates from the MOX zone to the UO_2 zone (thermal spectrum in UO_2 cell higher in the voided case than in the flooded situation): the neutrons coming from the UO_2 zone are driving the neutron spectrum in the L-MOX zone and consequently the spectrum is intermediate between the flooded situation and the fundamental mode which could be obtained in an infinite voided L-MOX lattice.

The analysis of the adjoint fluxes shows large modifications in the low thermal range where the voided L-MOX cell importance becomes higher than the others.

The comparison of the total importances has shown a same effect for the UO_2 cells, but a very large decrease in the thermal range for the L-MOX cells and consequently a very large increase in the fast energy range.

Analysis using Perturbations Theory

In order to improve our understanding of the physical phenomena occurring during the voidage of the L-MOX macrocell, we use perturbation calculation modules implemented in the ERANOS system.

We use the following macrocell model:

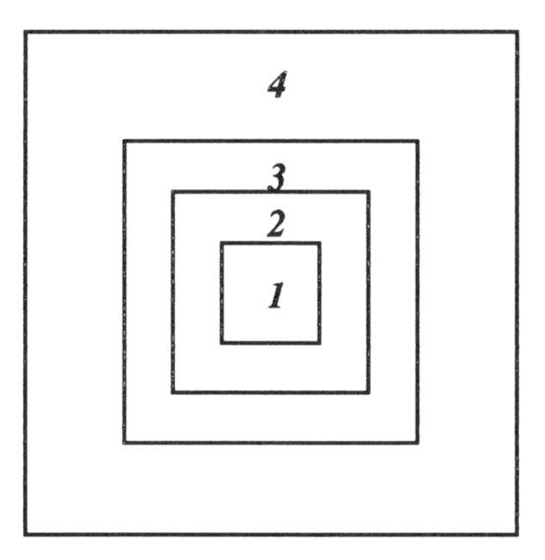

where
1. Central L-MOX zone
2. Peripheral L-MOX ring
3. First UO_2 ring
4. "Asymptotic" UO_2 zone

Calculations were done using the P_1, S_8 transport theory in direct and adjoint conditions in a 15 energy group structure. The cross-sections used were obtained by the ECCO code. However, the results of this study must be used with care because we have done *exact* perturbation calculations on the whole domain. Thus the contribution of each region cannot be derived from these results. In particular, it is impossible to separate correctly the balance between the MOX zone and the UO_2 zone.

Consequently, the conclusions we will give here must be used as qualitative information about the contributions of the macrogroups.

The contributions of each macrogroup to the total void reactivity are given in Table C.1-7.

The positive voidage effect is essentially due to a positive contribution of the capture reaction rate (positive effect) which is compensated by a negative contribution due to the elastic and inelastic slowing-down.

If we analyse the contribution of each macrogroup, we can see that the main positive contribution comes from macrogroups 2, 3, 4 and 5 (elastic and inelastic slowing-down) and the main negative effect is due to the macrogroup 12.

The very large positive effects of the capture (groups 11 and 13) are compensated by a negative contribution of the fissions (group 11) and of the slowing-down (groups 11, 12 and 13). In fact, the negative contribution of the slowing-down in group 13 is almost completely compensated by the positive effect of the capture.

For the group 15, we can see a cancellation between the capture (-107 pcm) and the fission (+153 pcm).

GROUP	ENERGY	CAPTURE	FISSION	ELASTIC + INELASTIC	TRANSPORT	TOTAL
1	19.6 MeV → 6.1 MeV	8.0	-1.5	16.1	0.0	22.8
2	6.1 MeV → 2.2 MeV	14.3	-7.1	127.4	0.0	134.5
3	2.2 MeV → 1.4 MeV	0.3	-0.5	105.7	0.0	100.6
4	1.4 MeV → 498 keV	1.4	-6.9	120.6	1.4	116.5
5	498 keV → 183 keV	0.7	-0.9	98.8	3.2	101.8
6	183 keV → 67 keV	0.5	-0.2	68.5	1.9	70.5
7	67 keV → 25 keV	1.6	0.0	22.0	0.5	24.0
8	25 keV → 9 keV	4.3	-0.1	-3.1	0.0	1.2
9	9 keV → 2 keV	23.4	-1.4	-33.5	-0.6	-12.1
10	2 keV → 454 eV	56.8	-7.2	-55.3	-1.2	-6.8
11	454 eV → 22.6 eV	360.6	-104.6	-279.6	-6.0	-29.6
12	22.6 eV → 4 eV	-63.9	-0.9	-220.1	-10.4	-295.2
13	4 eV → 0.53 eV	816.6	-65.0	-727.5	-7.6	16.6
14	0.53 eV → 0.1 eV	3.6	-16.9	-33.9	-8.7	-55.9
15	0.1 eV → 1.1 10^{-4} eV	-107.2	152.5	-7.2	-9.9	28.2
TOTAL	20 MeV → 1.1 10^{-4} eV	1121.2	-65.5	-801.1	-37.6	217.0

Table C.1-7 ***Contribution of the 15 macrogroups to the void reactivity effect pin the L-MOX macrocell (pcm)***

This study allows us to show that the total voidage effect comes from cancellations of positive and negative effects of each macrogroup used in the calculations; but, because we calculated exact perturbation effects, it was not possible to derive the contribution of the different zones of the macrocell. However, we can note a very large positive effect of the capture rate (i.e., reduced capture) which is compensated by a strong negative effect of the elastic and inelastic slowing-down towards the low energy domain. The main positive part comes from the removal in the high energy range.

These contributions are related to the spectrum of the neutron importance, which decreases with energy at high energy and increases at low energy.

Conclusion

During the first phase of the benchmark study the participants observed a positive voidage effect for the macrocells loaded with MOX fuel, even for the lowest plutonium content. It is the reason why a special task has been performed at CEA to try to understand the physical phenomena which could explain this situation.

Several studies have been performed using the APOLLO-2 code (CEA-93 library – JEF-2.2 data) and the ERANOS system (ECCO code + JEF-2 data and perturbation models).

The main conclusion to be drawn from these studies is that the positive voidage reactivity coefficient is a physical phenomenon (spectral effect) related to the specifications of the benchmark. It is clear that in realistic cases, we must also take into account the negative contribution of the leakage which might compensate the positive spectral effect.

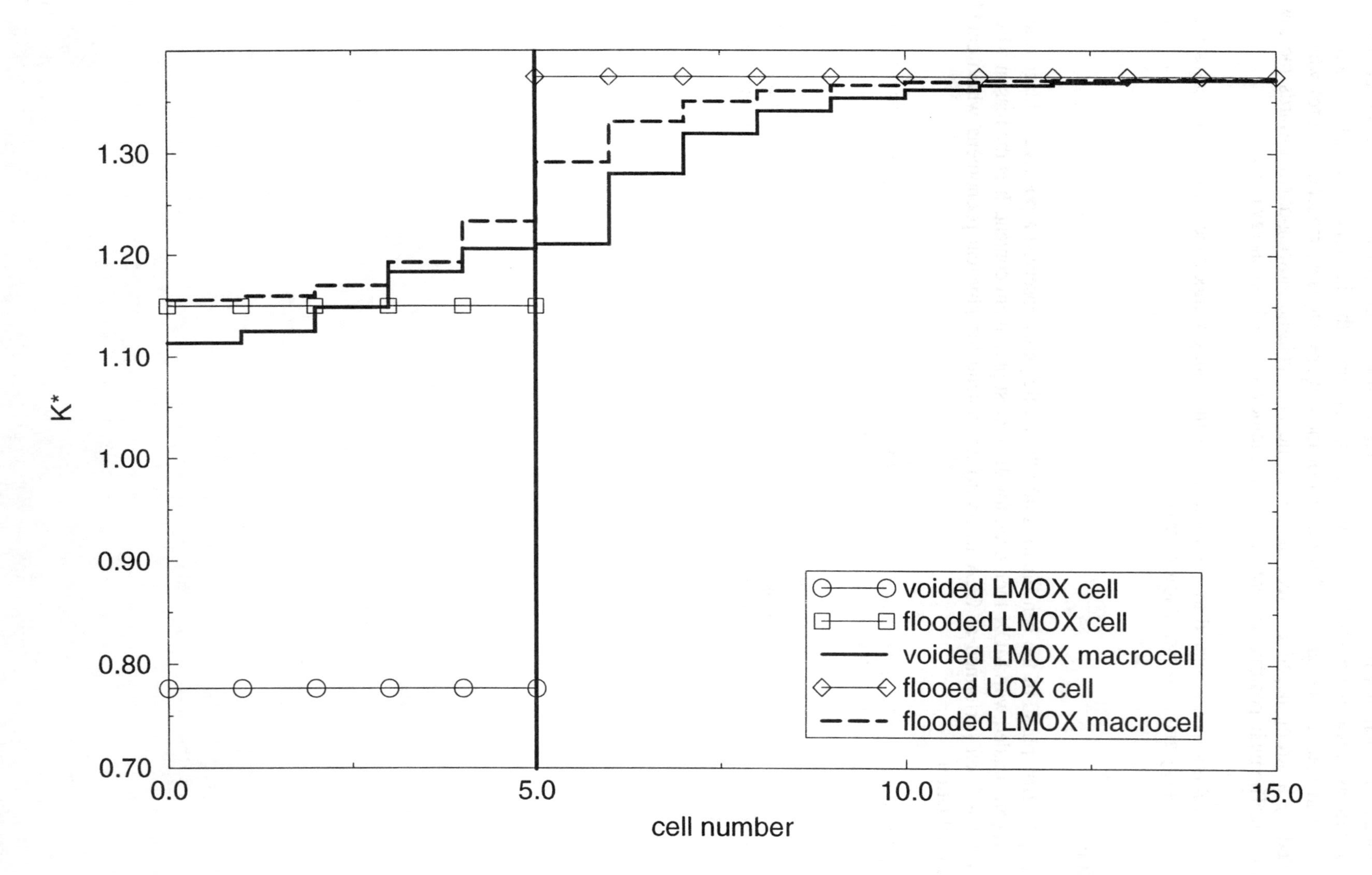

Figure C.1-1 ***L-MOX macrocell – variation of k* along AM axis***

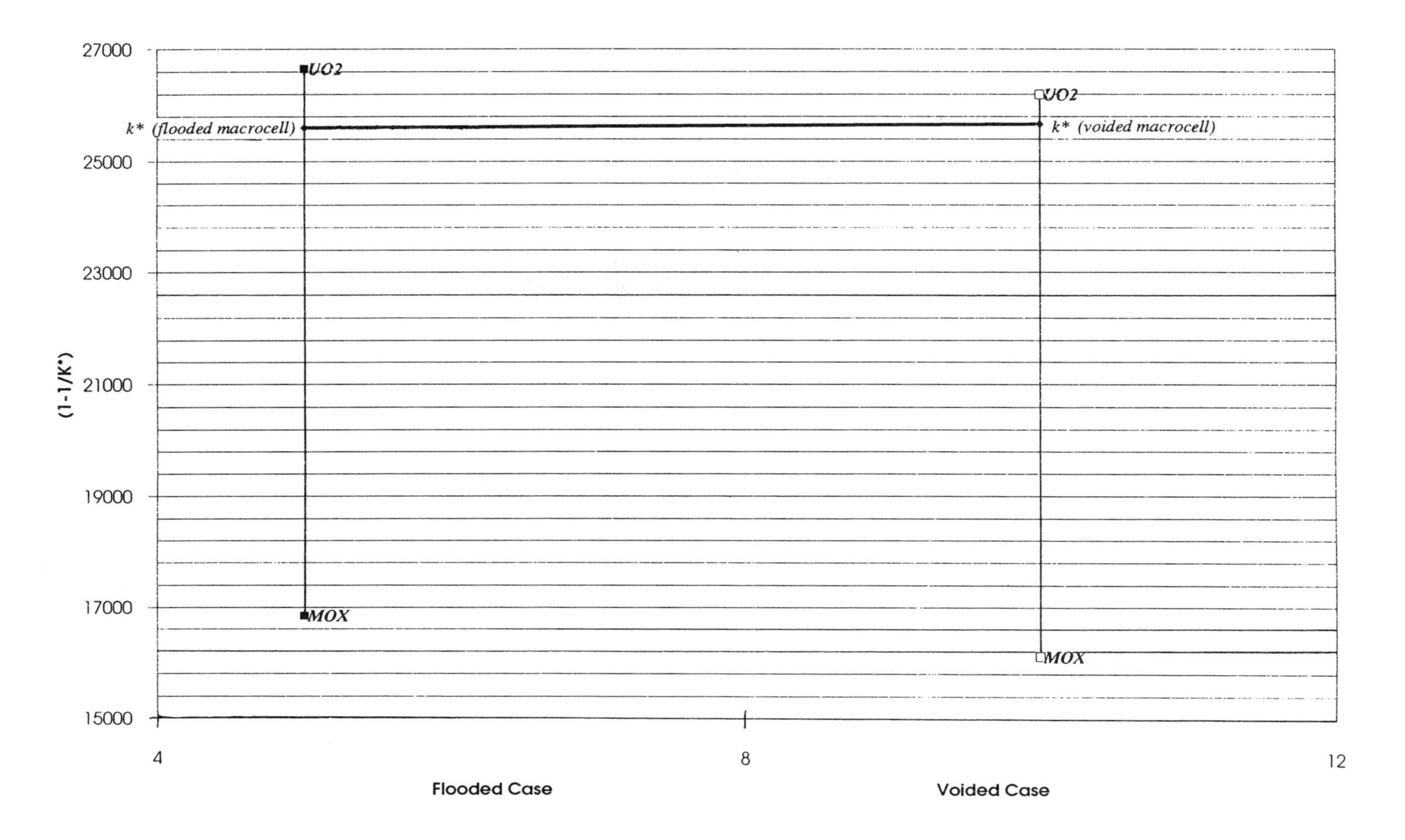

Figure C.1-2 ***Comparison of the reactivity for flooded and voided situations***

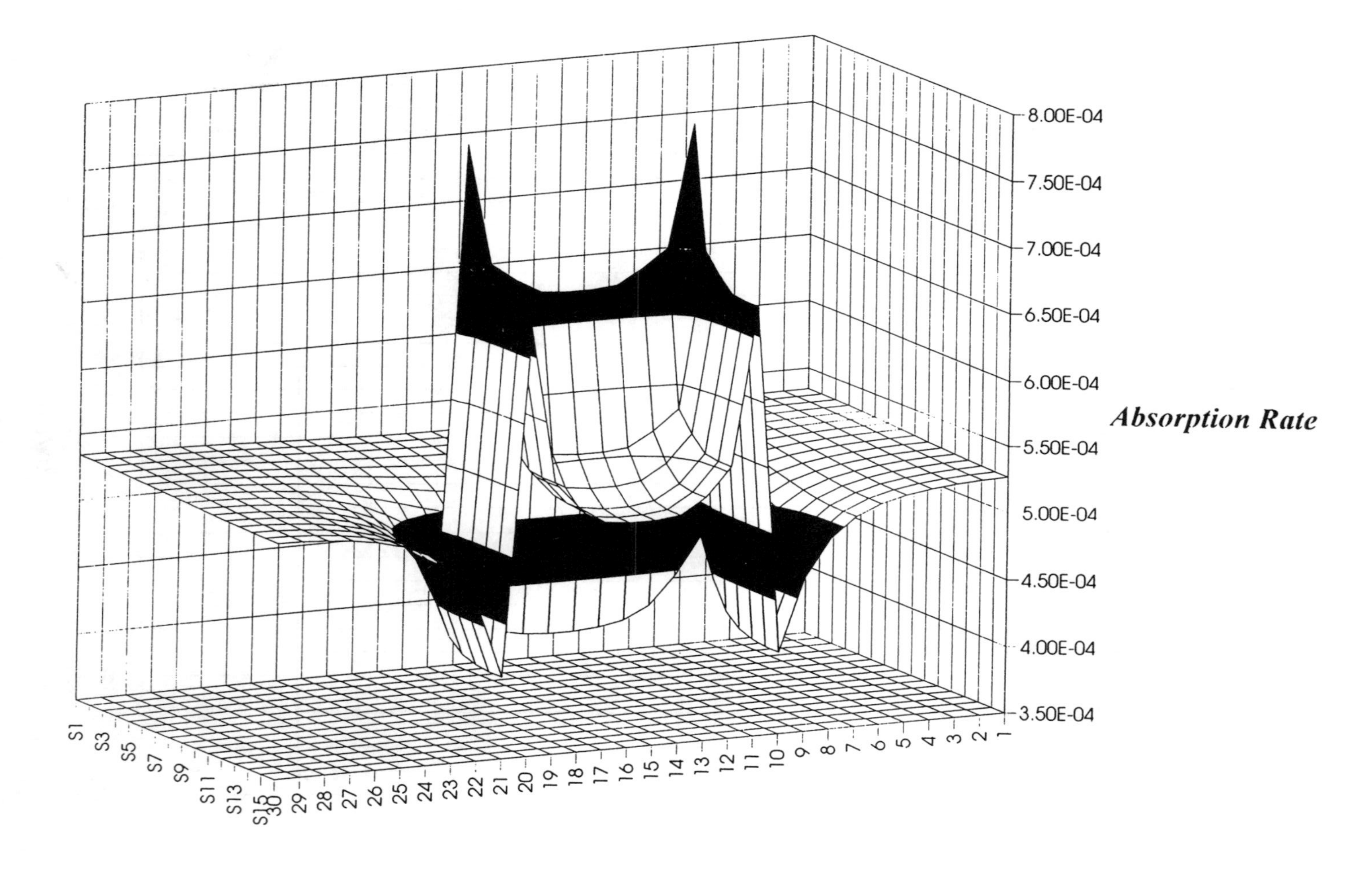

Figure C.1-3 ***Absorption rate in the flooded L-MOX macrocell***

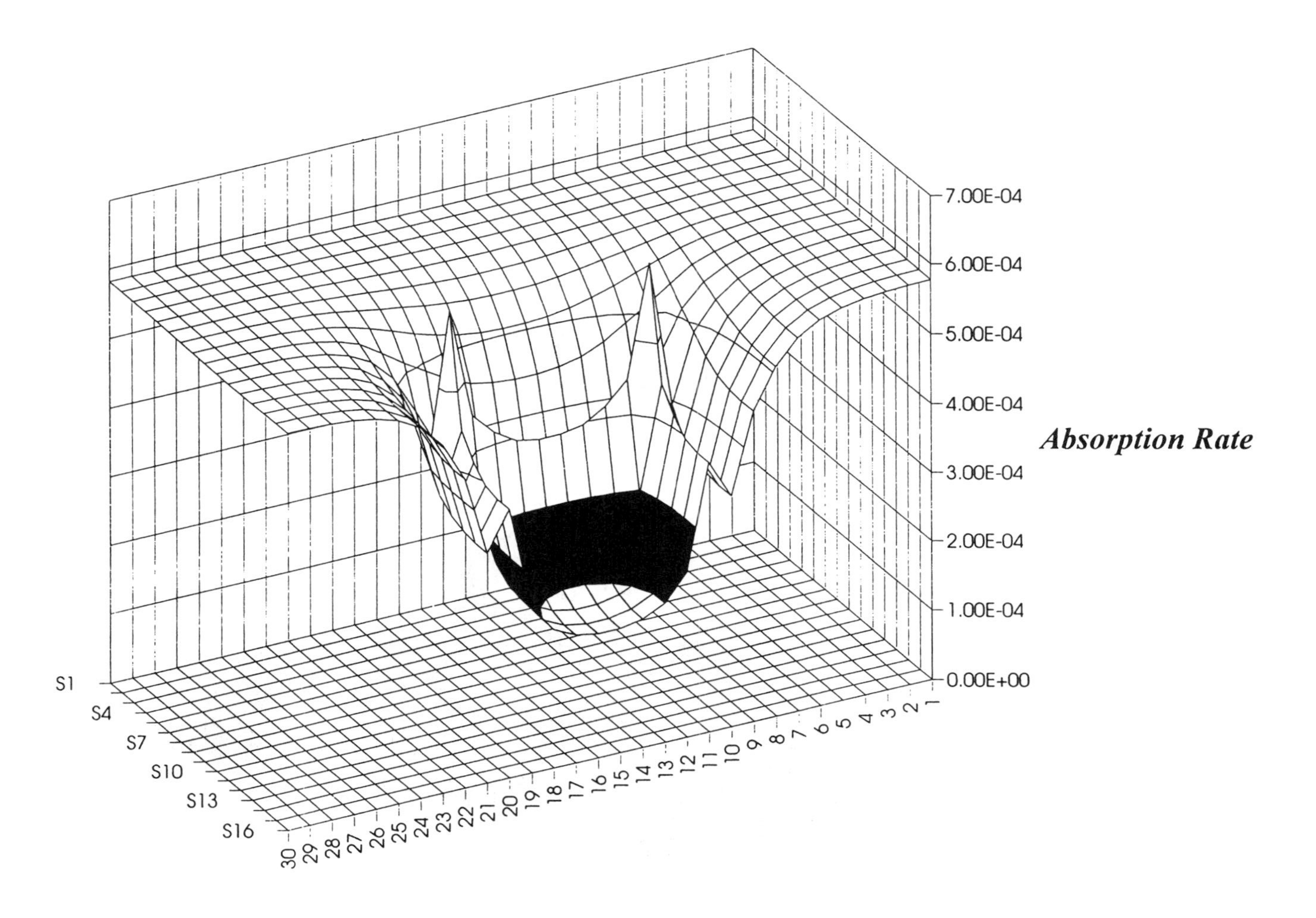

Figure C.1-4 ***Absorption rate in the voided L-MOX macrocell***

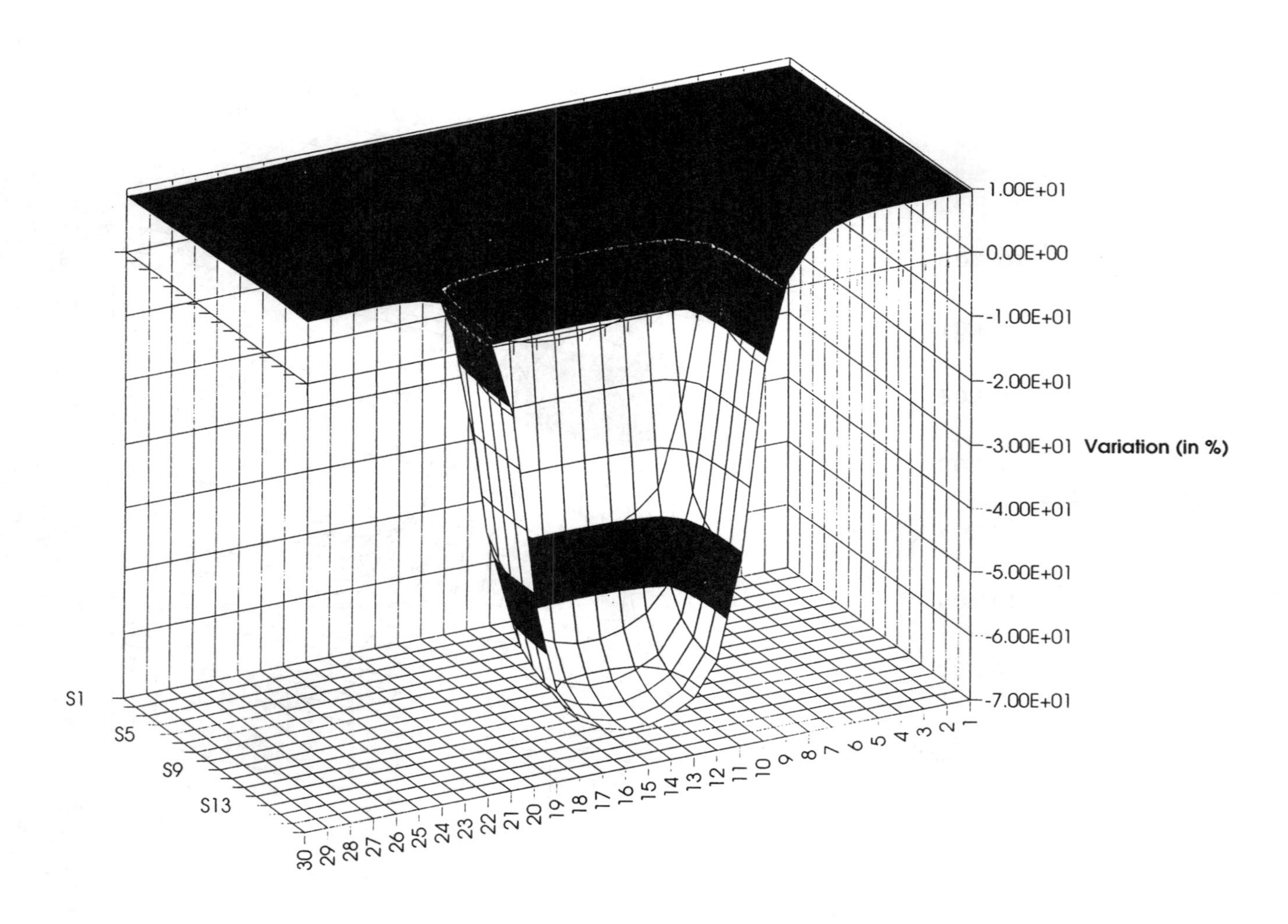

Figure C.1-5 ***Variation of the capture rate***

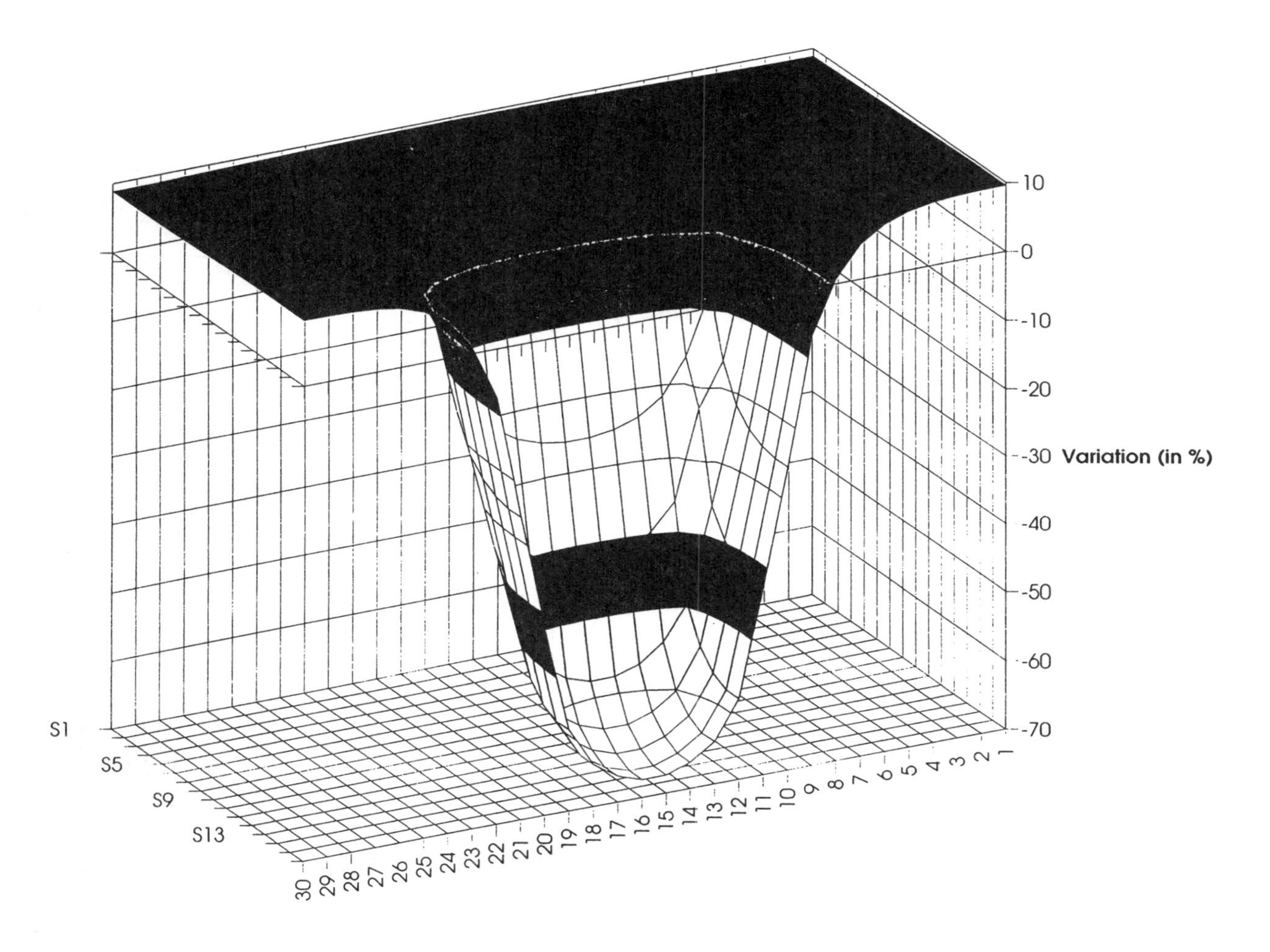

Figure C.1-6 ***Variation of the fission rate***

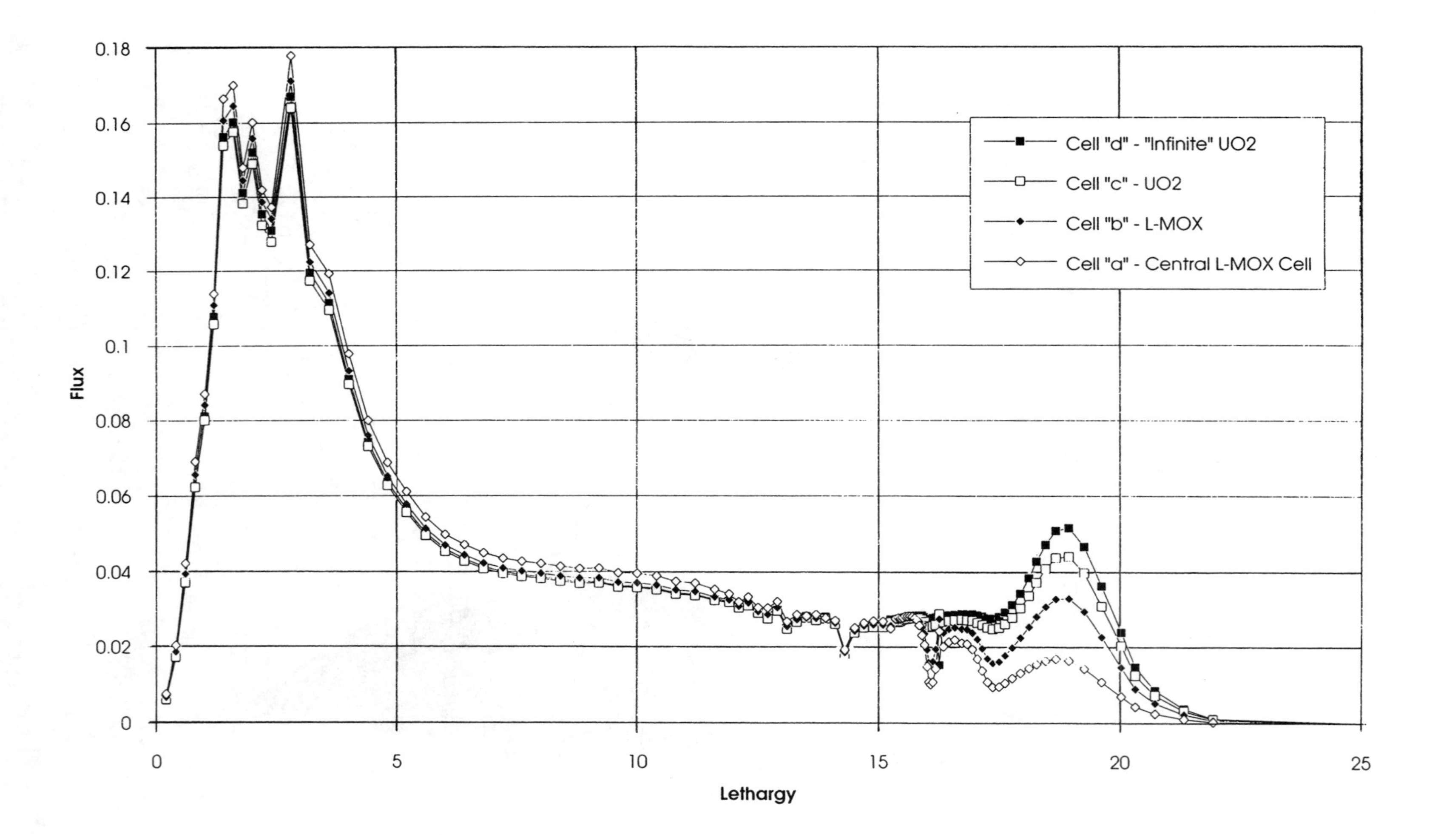

Figure C.1-7 ***Direct flux in the flooded L-MOX macrocell***

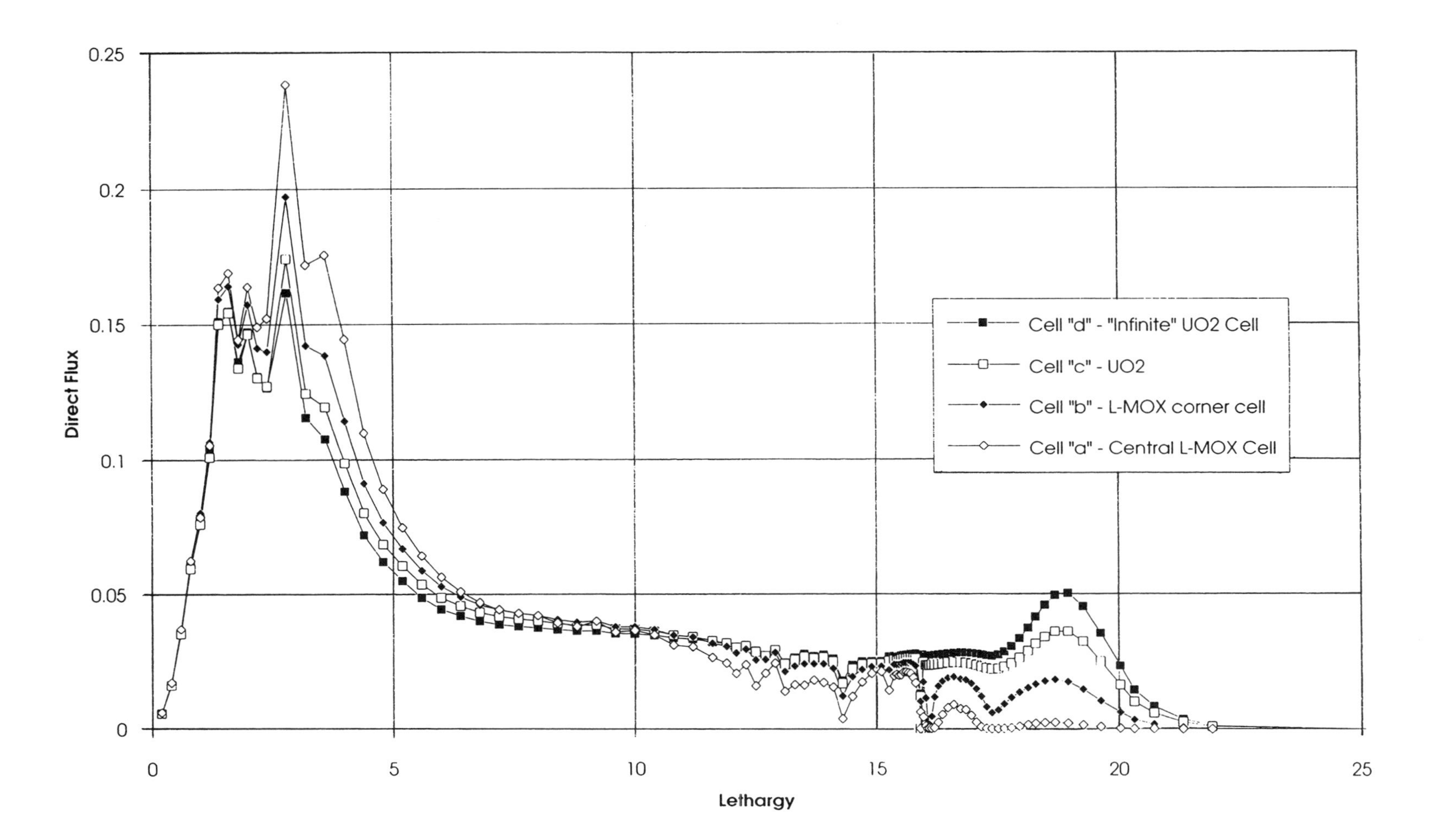

Figure C.1-8 ***Direct flux in the voided L-MOX macrocell***

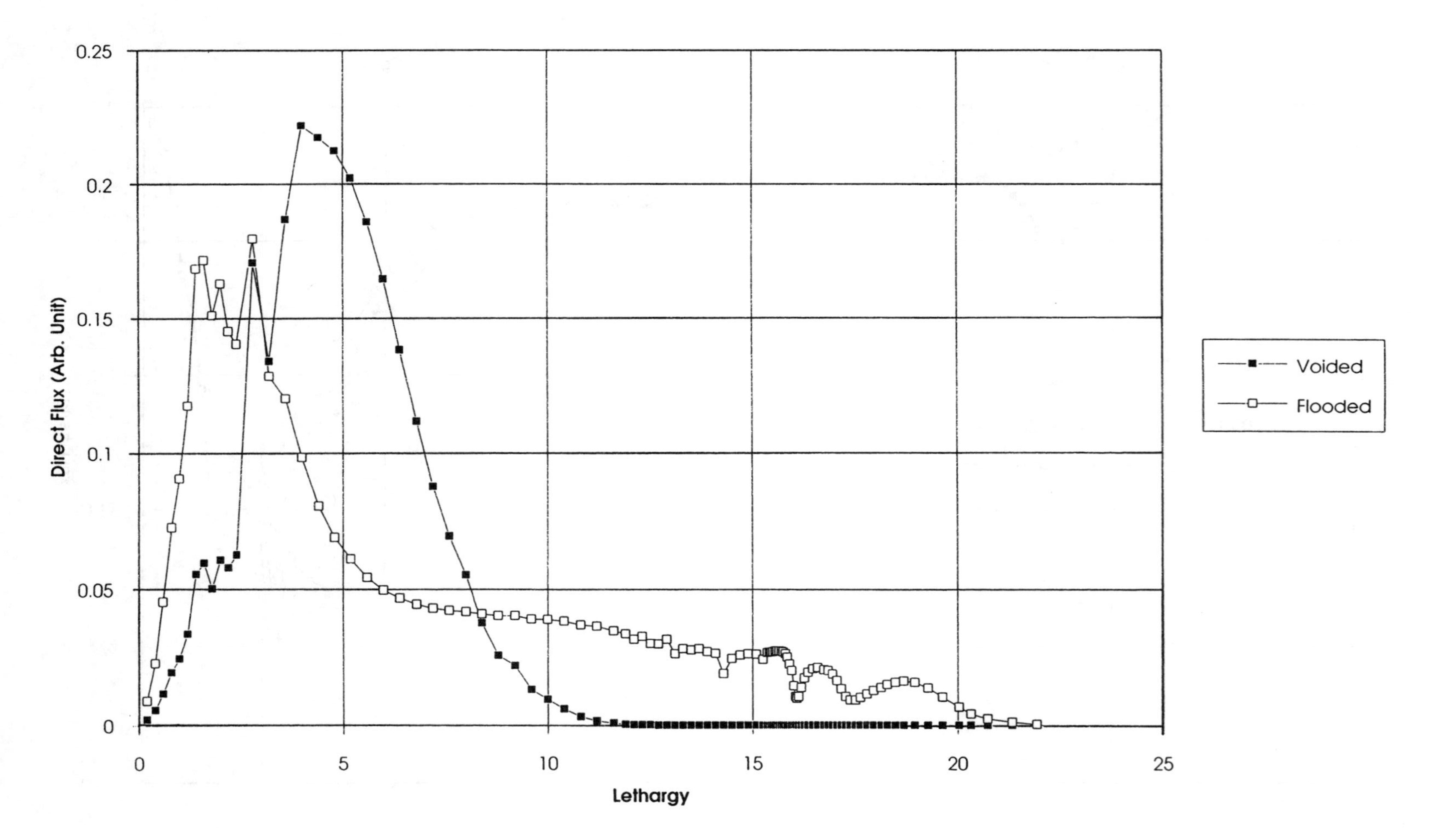

Figure C.1-9 ***L-MOX infinite medium — direct flux***

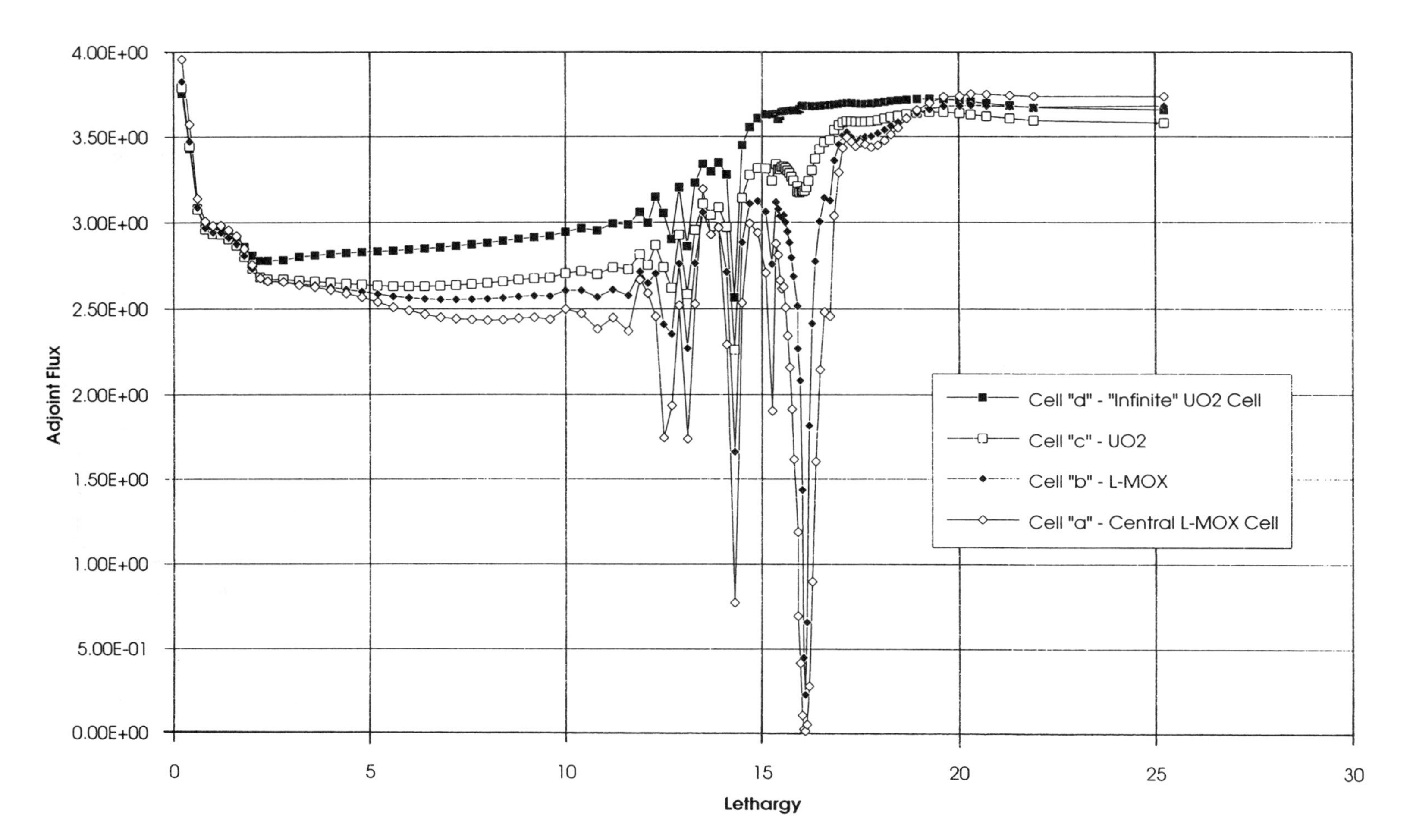

Figure C.1-10 ***Adjoint flux in the voided L-MOX macrocell***

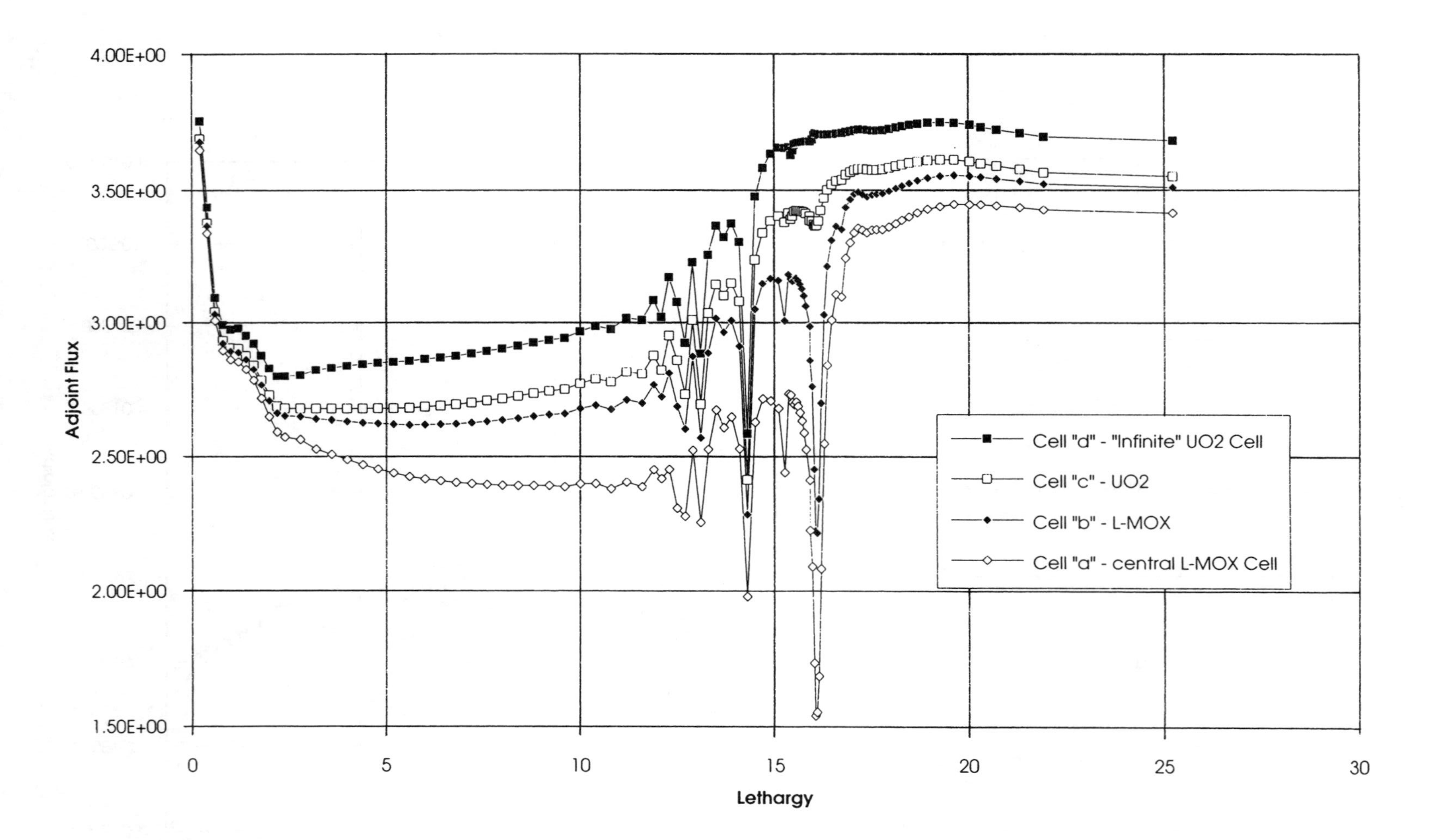

Figure C.1-11 ***Adjoint in the flooded L-MOX macrocell***

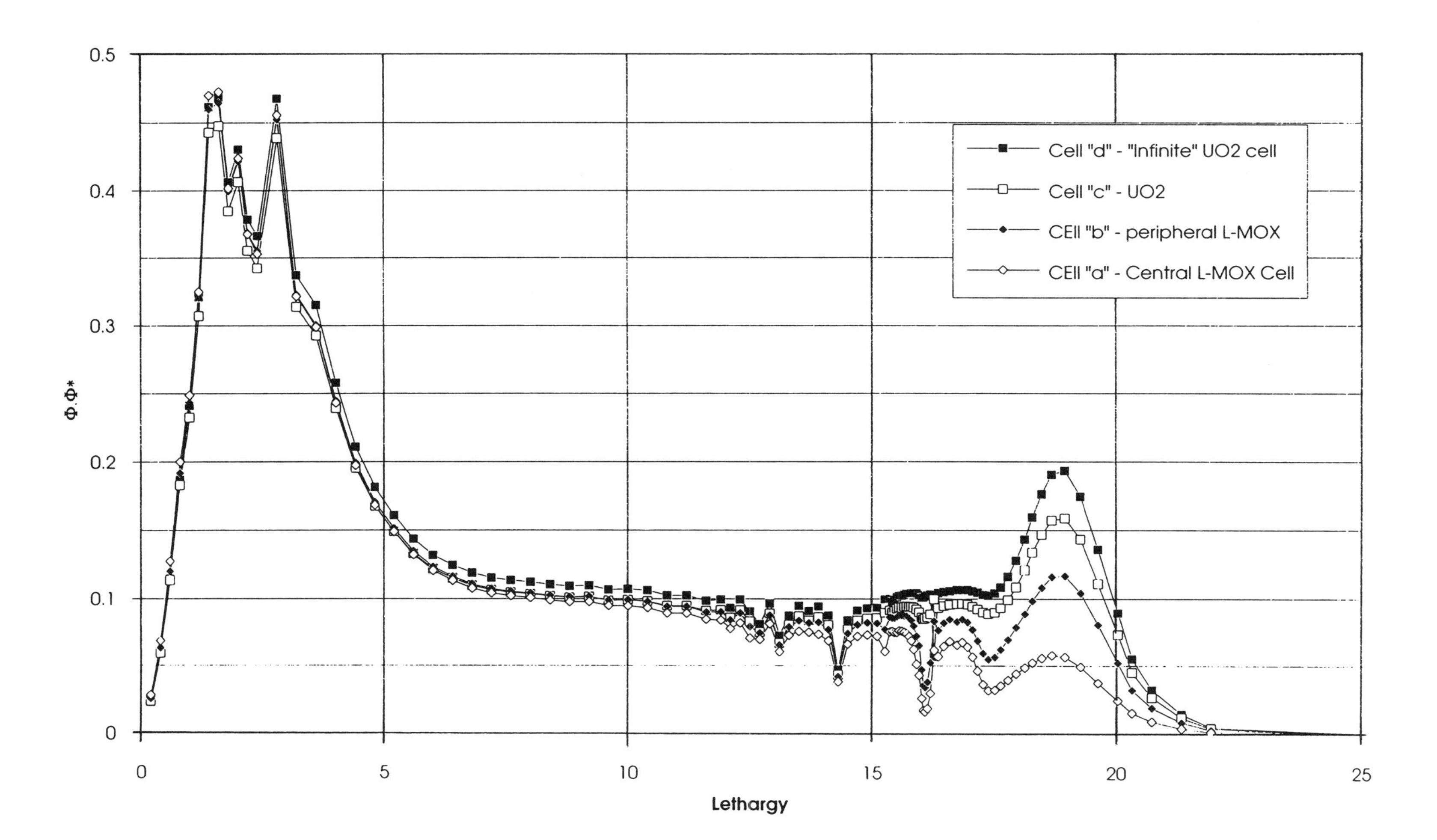

Figure C.1-12 ***Total importance for the flooded L-MOX macrocell***

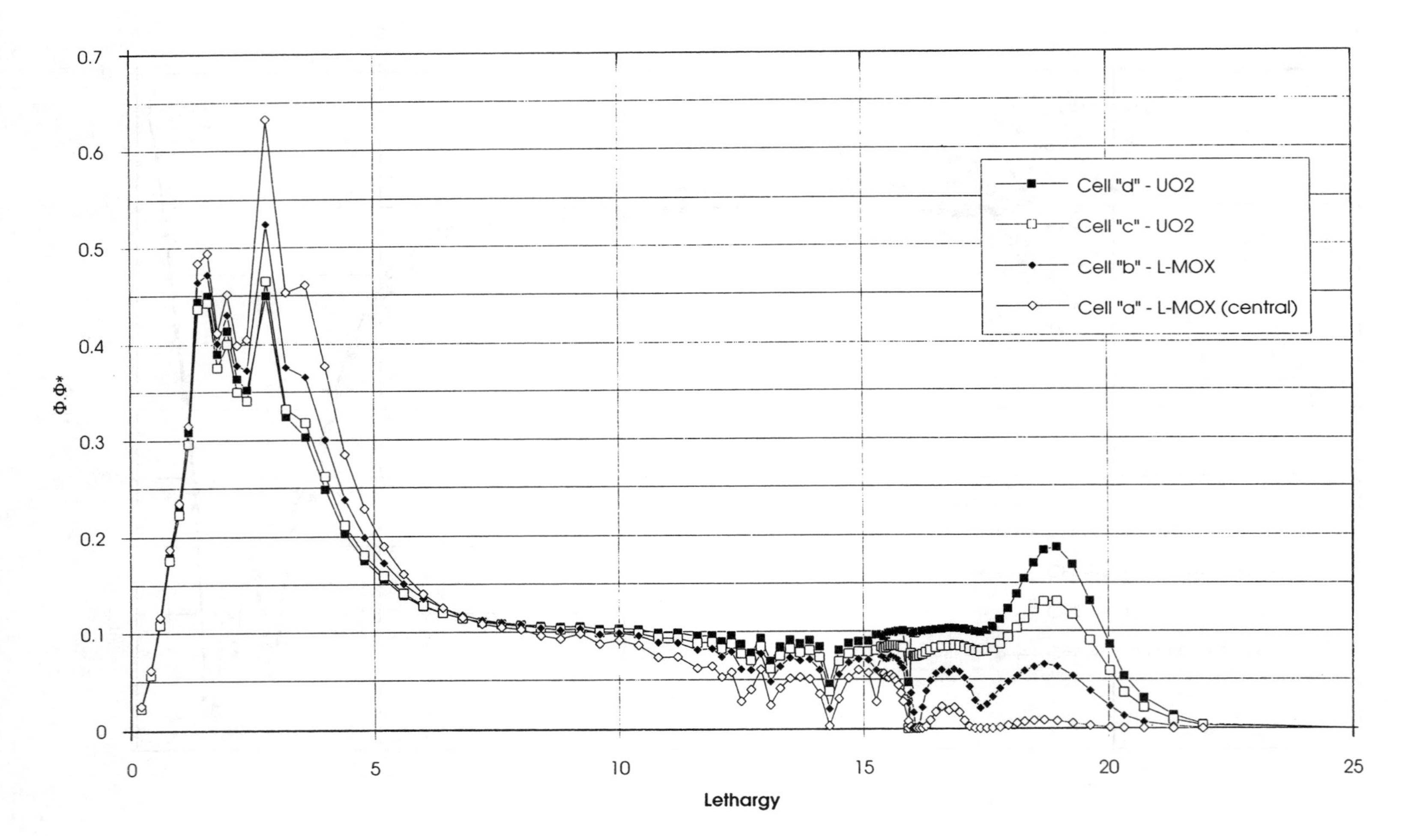

Figure C.1-13 ***Total importance for the voided L-MOX macrocell***

Appendix C.2

Comparison of the results calculated with several Monte Carlo codes and nuclear data for plutonium recycling and void reactivity effect benchmark in PWRs

H. Takano, T. Mori, H. Akie (JAERI)

Comparison of the calculated with continuous energy Monte Carlo codes: VIM, MVP & MCNP

Benchmark cell	**H-MOX**	
Nuclear data	**JENDL-3.1**	
Cross-section library	**VIMLIB**	unresolved region – probability table pointwise cross section – 0.1% error
	MVPLIB	unresolved region – probability table / infinite option pointwise cross-section – 0.1% error
	MCNPLIB	unresolved region – infinite diluted pointwise cross-section – 0.5% error
	MVPMCNPLIB: MCNPLIB	transferred to MVP code format

- The comparison between VIM and MVP shows a very good agreement between both results for the unvoided and voided cell calculations;
- The results calculated with MCNP-4.2 and MVP (MVPMCNPLIB) are in good agreement within the statistical errors of MCNP-4.2;
- The resonance shielding effect in the unresolved region are 0.0005 Δk for the unvoided and 0.0080 Δk for the voided cell calculations from comparing between the Zr-isotopes cases for MVPLIB and MVPMCNPLIB. And the effect is 0.0082 Δk for voided Zr-natural case of MVPLIB;

- The discrepancies between Zr-natural and isotope-wise cross-sections show the following effect on k-effective:

  – unvoided case: very small,
  – voided case: probability table $\Delta k = k_{iso} - k_{nat} =$ 0.0021
  infinite $\Delta k =$ 0.0013
  MCNP-4.2 $\Delta k =$ 0.0037

- The difference between MCNP-4.2–JAERI and MCNP-4.2–Toshiba is 0.014 Δk for the voided case.

Table C.2-1 ***Comparison of the results calculated with the different Monte Carlo codes for H-MOX cell model using the cross-section library based on JENDL-3.1.***

	VIM VIMLIB	**MVP MVPLIB**	**MVP MVPMCNPLIB**	**MCNP-4.2 MCNPLIB**	**MCNP-4.2 MCNPLIB / Toshiba**
unvoided Zr-natural	1.2161 ± 0.05 (prob. table)	1.2161 ± 0.03 (prob. table)			
unvoided Zr-isotopes		1.2155 ± 0.03 (prob. table)	1.2150 ± 0.02 (infinite)	1.2166 ± 0.15 (infinite)	1.2165 ± 0.09 (infinite)
voided Zr-natural	1.2696 ± 0.05 (prob. table)	1.2694 ± 0.02 (prob. table)			
		1.2612 ± 0.02 (infinite)	1.2622 ± 0.03 (infinite)	1.2610 ± 0.13 (infinite)	
unvoided Zr-isotopes		1.2715 ± 0.02 (prob. table)	1.2635 ± 0.03 (infinite)	1.2647 ± 0.12 (infinite)	1.2783 ± 0.10 (infinite)

Comparison of the results calculated with JENDL-3.1 and 3.2 nuclear data

- The Monte Carlo code MVP was revised for upscattering treatment by expanding from 2 to 4 eV using the short collision scattering law. And the results recalculated for the void cell benchmark show about 0.3% Δk effect for unvoided infinite cell calculations though they are very small for voided cell and assembly cases;

- The results calculated with JENDL-3.2 are about 1% Δk larger than those for JENDL-3.1 for UO_2 cell and H-MOX assembly cases. This is due to the capture resonance integral of U-235 for JENDL-3.2 is 13% larger than that of JENDL-3.1. The difference between JENDL-3.1 and 3.2 is small for the H-MOX calculation.

Table C.2-2 ***Values corrected for MVP-JENDL-3.1 and comparison of the results calculated by using nuclear data JENDL-3.1 and 3.2.***

	NUCLEAR DATA	UO_2	H-MOX	M-MOX	L-MOX
infinite lattice	JENDL-3.1 4 eV 2 eV	1.3628 (1.3622)	1.2157 (1.2185)	1.1722 (1.1757)	1.1507 (1.1533)
unvoided	JENDL-3.2	1.3728	1.2168		
infinite lattice	JENDL-3.1 4 eV 2 eV	0.6221 (0.6228)	1.2696 (1.2689)	1.0284 (1.0280)	0.7579 (0.7595)
voided	JENDL-3.2	0.6180	1.2679		
assembly	JENDL-3.1 4 eV 2 eV	1.3624 (1.3629)	1.3414 (1.3416)	1.3371 (1.3383)	1.3367 (1.3374)
unvoided	JENDL-3.2	1.3728	1.3499		
assembly	JENDL-3.1 4 eV 2 eV	1.3474 (1.3474)	1.3453 (1.3457)	1.3405 (1.3415)	1.3372 (1.3372)
voided	JENDL-3.2	1.3580	1.3550		

(...) show the old values reported previously with upscattering limited to 2 eV.

List of symbols and abbreviations

at/10^{-24} cm^3	atomic density
°C	degrees Celsius
c/a	*c*alculated *a*verage ratio
contr.	contributor
CPU	central processor unit
Δk	variation of multiplication factor
Δu	lethargy width
η	number of neutrons released per neutron absorbed
eV	electronvolt
k	neutron multiplication factor
keV	kiloelectronvolt
MOX **L-MOX** **M-MOX** **H-MOX**	*M*ixed *O*xide (uranium and plutonium) MOX with *low* plutonium content MOX with *medium* plutonium content MOX with *high* plutonium content
ν	neutron per fission
NEM	*N*odal *E*xpansion *M*ethod
OECD/NEA	OECD Nuclear Energy Agency
pcm	10^{-5}
PT	*P*erturbation *T*heory
Pu	*plutonium*
PWR	*P*ressurized-*W*ater *R*eactor
σ	standard deviation
S_n	discrete ordinates radiation transport method
U	uranium
UO_2	uranium dioxide = **UOX**
w/o	weight % = wt%
WPPR	*W*orking Party on the *P*hysics of *P*lutonium *R*ecycling

MAIN SALES OUTLETS OF OECD PUBLICATIONS
PRINCIPAUX POINTS DE VENTE DES PUBLICATIONS DE L'OCDE

ARGENTINA – ARGENTINE
Carlos Hirsch S.R.L.
Galería Güemes, Florida 165, 4° Piso
1333 Buenos Aires Tel. (1) 331.1787 y 331.2391
Telefax: (1) 331.1787

AUSTRALIA – AUSTRALIE
D.A. Information Services
648 Whitehorse Road, P.O.B 163
Mitcham, Victoria 3132 Tel. (03) 873.4411
Telefax: (03) 873.5679

AUSTRIA – AUTRICHE
Gerold & Co.
Graben 31
Wien I Tel. (0222) 533.50.14
Telefax: (0222) 512.47.31.29

BELGIUM – BELGIQUE
Jean De Lannoy
Avenue du Roi 202 Koningslaan
B-1060 Bruxelles Tel. (02) 538.51.69/538.08.41
Telefax: (02) 538.08.41

CANADA
Renouf Publishing Company Ltd.
1294 Algoma Road
Ottawa, ON K1B 3W8 Tel. (613) 741.4333
Telefax: (613) 741.5439
Stores:
61 Sparks Street
Ottawa, ON K1P 5R1 Tel. (613) 238.8985
211 Yonge Street
Toronto, ON M5B 1M4 Tel. (416) 363.3171
Telefax: (416)363.59.63

Les Éditions La Liberté Inc.
3020 Chemin Sainte-Foy
Sainte-Foy, PQ G1X 3V6 Tel. (418) 658.3763
Telefax: (418) 658.3763

Federal Publications Inc.
165 University Avenue, Suite 701
Toronto, ON M5H 3B8 Tel. (416) 860.1611
Telefax: (416) 860.1608

Les Publications Fédérales
1185 Université
Montréal, QC H3B 3A7 Tel. (514) 954.1633
Telefax: (514) 954.1635

CHINA – CHINE
China National Publications Import
Export Corporation (CNPIEC)
16 Gongti E. Road, Chaoyang District
P.O. Box 88 or 50
Beijing 100704 PR Tel. (01) 506.6688
Telefax: (01) 506.3101

CHINESE TAIPEI – TAIPEI CHINOIS
Good Faith Worldwide Int'l. Co. Ltd.
9th Floor, No. 118, Sec. 2
Chung Hsiao E. Road
Taipei Tel. (02) 391.7396/391.7397
Telefax: (02) 394.9176

CZECH REPUBLIC – RÉPUBLIQUE TCHEQUE
Artia Pegas Press Ltd.
Narodni Trida 25
POB 825
111 21 Praha 1 Tel. 26.65.68
Telefax: 26.20.81

DENMARK – DANEMARK
Munksgaard Book and Subscription Service
35, Nørre Søgade, P.O. Box 2148
DK-1016 København K Tel. (33) 12.85.70
Telefax: (33) 12.93.87

EGYPT – ÉGYPTE
Middle East Observer
41 Sherif Street
Cairo Tel. 392.6919
Telefax: 360-6804

FINLAND – FINLANDE
Akateeminen Kirjakauppa
Keskuskatu 1, P.O. Box 128
00100 Helsinki
Subscription Services/Agence d'abonnements :
P.O. Box 23
00371 Helsinki Tel. (358 0) 121 4416
Telefax: (358 0) 121.4450

FRANCE
OECD/OCDE
Mail Orders/Commandes par correspondance:
2, rue André-Pascal
75775 Paris Cedex 16 Tel. (33-1) 45.24.82.00
Telefax: (33-1) 49.10.42.76
Telex: 640048 OCDE
Internet: Compte.PUBSINQ @ oecd.org
Orders via Minitel, France only/
Commandes par Minitel, France exclusivement :
36 15 OCDE
OECD Bookshop/Librairie de l'OCDE :
33, rue Octave-Feuillet
75016 Paris Tel. (33-1) 45.24.81.81
(33-1) 45.24.81.67

Documentation Française
29, quai Voltaire
75007 Paris Tel. 40.15.70.00
Gibert Jeune (Droit-Économie)
6, place Saint-Michel
75006 Paris Tel. 43.25.91.19
Librairie du Commerce International
10, avenue d'Iéna
75016 Paris Tel. 40.73.34.60
Librairie Dunod
Université Paris-Dauphine
Place du Maréchal de Lattre de Tassigny
75016 Paris Tel. (1) 44.05.40.13
Librairie Lavoisier
11, rue Lavoisier
75008 Paris Tel. 42.65.39.95
Librairie L.G.D.J. - Montchrestien
20, rue Soufflot
75005 Paris Tel. 46.33.89.85
Librairie des Sciences Politiques
30, rue Saint-Guillaume
75007 Paris Tel. 45.48.36.02
P.U.F.
49, boulevard Saint-Michel
75005 Paris Tel. 43.25.83.40
Librairie de l'Université
12a, rue Nazareth
13100 Aix-en-Provence Tel. (16) 42.26.18.08
Documentation Française
165, rue Garibaldi
69003 Lyon Tel. (16) 78.63.32.23
Librairie Decitre
29, place Bellecour
69002 Lyon Tel. (16) 72.40.54.54
Librairie Sauramps
Le Triangle
34967 Montpellier Cedex 2 Tel. (16) 67.58.85.15
Tekefax: (16) 67.58.27.36

GERMANY – ALLEMAGNE
OECD Publications and Information Centre
August-Bebel-Allee 6
D-53175 Bonn Tel. (0228) 959.120
Telefax: (0228) 959.12.17

GREECE – GRÈCE
Librairie Kauffmann
Mavrokordatou 9
106 78 Athens Tel. (01) 32.55.321
Telefax: (01) 32.30.320

HONG-KONG
Swindon Book Co. Ltd.
Astoria Bldg. 3F
34 Ashley Road, Tsimshatsui
Kowloon, Hong Kong Tel. 2376.2062
Telefax: 2376.0685

HUNGARY – HONGRIE
Euro Info Service
Margitsziget, Európa Ház
1138 Budapest Tel. (1) 111.62.16
Telefax: (1) 111.60.61

ICELAND – ISLANDE
Mál Mog Menning
Laugavegi 18, Pósthólf 392
121 Reykjavik Tel. (1) 552.4240
Telefax: (1) 562.3523

INDIA – INDE
Oxford Book and Stationery Co.
Scindia House
New Delhi 110001 Tel. (11) 331.5896/5308
Telefax: (11) 332.5993
17 Park Street
Calcutta 700016 Tel. 240832

INDONESIA – INDONÉSIE
Pdii-Lipi
P.O. Box 4298
Jakarta 12042 Tel. (21) 573.34.67
Telefax: (21) 573.34.67

IRELAND – IRLANDE
Government Supplies Agency
Publications Section
4/5 Harcourt Road
Dublin 2 Tel. 661.31.11
Telefax: 475.27.60

ISRAEL
Praedicta
5 Shatner Street
P.O. Box 34030
Jerusalem 91430 Tel. (2) 52.84.90/1/2
Telefax: (2) 52.84.93

R.O.Y. International
P.O. Box 13056
Tel Aviv 61130 Tel. (3) 546 1423
Telefax: (3) 546 1442

Palestinian Authority/Middle East:
INDEX Information Services
P.O.B. 19502
Jerusalem Tel. (2) 27.12.19
Telefax: (2) 27.16.34

ITALY – ITALIE
Libreria Commissionaria Sansoni
Via Duca di Calabria 1/1
50125 Firenze Tel. (055) 64.54.15
Telefax: (055) 64.12.57
Via Bartolini 29
20155 Milano Tel. (02) 36.50.83
Editrice e Libreria Herder
Piazza Montecitorio 120
00186 Roma Tel. 679.46.28
Telefax: 678.47.51

Libreria Hoepli
Via Hoepli 5
20121 Milano Tel. (02) 86.54.46
Telefax: (02) 805.28.86

Libreria Scientifica
Dott. Lucio de Biasio 'Aeiou'
Via Coronelli, 6
20146 Milano Tel. (02) 48.95.45.52
Telefax: (02) 48.95.45.48

JAPAN – JAPON
OECD Publications and Information Centre
Landic Akasaka Building
2-3-4 Akasaka, Minato-ku
Tokyo 107 Tel. (81.3) 3586.2016
Telefax: (81.3) 3584.7929

KOREA – CORÉE
Kyobo Book Centre Co. Ltd.
P.O. Box 1658, Kwang Hwa Moon
Seoul Tel. 730.78.91
Telefax: 735.00.30

MALAYSIA – MALAISIE
University of Malaya Bookshop
University of Malaya
P.O. Box 1127, Jalan Pantai Baru
59700 Kuala Lumpur
Malaysia Tel. 756.5000/756.5425
Telefax: 756.3246

MEXICO – MEXIQUE
Revistas y Periodicos Internacionales S.A. de C.V.
Florencia 57 - 1004
Mexico, D.F. 06600 Tel. 207.81.00
Telefax: 208.39.79

NETHERLANDS – PAYS-BAS
SDU Uitgeverij Plantijnstraat
Externe Fondsen
Postbus 20014
2500 EA's-Gravenhage Tel. (070) 37.89.880
Voor bestellingen: Telefax: (070) 34.75.778

NEW ZEALAND NOUVELLE-ZÉLANDE
GPLegislation Services
P.O. Box 12418
Thorndon, Wellington Tel. (04) 496.5655
Telefax: (04) 496.5698

NORWAY – NORVÈGE
Narvesen Info Center – NIC
Bertrand Narvesens vei 2
P.O. Box 6125 Etterstad
0602 Oslo 6 Tel. (022) 57.33.00
Telefax: (022) 68.19.01

PAKISTAN
Mirza Book Agency
65 Shahrah Quaid-E-Azam
Lahore 54000 Tel. (42) 353.601
Telefax: (42) 231.730

PHILIPPINE – PHILIPPINES
International Book Center
5th Floor, Filipinas Life Bldg.
Ayala Avenue
Metro Manila Tel. 81.96.76
Telex 23312 RHP PH

PORTUGAL
Livraria Portugal
Rua do Carmo 70-74
Apart. 2681
1200 Lisboa Tel. (01) 347.49.82/5
Telefax: (01) 347.02.64

SINGAPORE – SINGAPOUR
Gower Asia Pacific Pte Ltd.
Golden Wheel Building
41, Kallang Pudding Road, No. 04-03
Singapore 1334 Tel. 741.5166
Telefax: 742.9356

SPAIN – ESPAGNE
Mundi-Prensa Libros S.A.
Castelló 37, Apartado 1223
Madrid 28001 Tel. (91) 431.33.99
Telefax: (91) 575.39.98

Libreria Internacional AEDOS
Consejo de Ciento 391
08009 – Barcelona Tel. (93) 488.30.09
Telefax: (93) 487.76.59

Llibreria de la Generalitat
Palau Moja
Rambla dels Estudis, 118
08002 – Barcelona
(Subscripcions) Tel. (93) 318.80.12
(Publicacions) Tel. (93) 302.67.23
Telefax: (93) 412.18.54

SRI LANKA
Centre for Policy Research
c/o Colombo Agencies Ltd.
No. 300-304, Galle Road
Colombo 3 Tel. (1) 574240, 573551-2
Telefax: (1) 575394, 510711

SWEDEN – SUÈDE
Fritzes Customer Service
S–106 47 Stockholm Tel. (08) 690.90.90
Telefax: (08) 20.50.21

Subscription Agency/Agence d'abonnements :
Wennergren-Williams Info AB
P.O. Box 1305
171 25 Solna Tel. (08) 705.97.50
Telefax: (08) 27.00.71

SWITZERLAND – SUISSE
Maditec S.A. (Books and Periodicals - Livres et périodiques)
Chemin des Palettes 4
Case postale 266
1020 Renens VD 1 Tel. (021) 635.08.65
Telefax: (021) 635.07.80

Librairie Payot S.A.
4, place Pépinet
CP 3212
1002 Lausanne Tel. (021) 341.33.47
Telefax: (021) 341.33.45

Librairie Unilivres
6, rue de Candolle
1205 Genève Tel. (022) 320.26.23
Telefax: (022) 329.73.18

Subscription Agency/Agence d'abonnements :
Dynapresse Marketing S.A.
38 avenue Vibert
1227 Carouge Tel. (022) 308.07.89
Telefax: (022) 308.07.99

See also – Voir aussi :
OECD Publications and Information Centre
August-Bebel-Allee 6
D-53175 Bonn (Germany) Tel. (0228) 959.120
Telefax: (0228) 959.12.17

THAILAND – THAÏLANDE
Suksit Siam Co. Ltd.
113, 115 Fuang Nakhon Rd.
Opp. Wat Rajbopith
Bangkok 10200 Tel. (662) 225.9531/2
Telefax: (662) 222.5188

TURKEY – TURQUIE
Kültür Yayinlari Is-Türk Ltd. Sti.
Atatürk Bulvari No. 191/Kat 13
Kavaklidere/Ankara Tel. 428.11.40 Ext. 2458
Dolmabahce Cad. No. 29
Besiktas/Istanbul Tel. (312) 260 7188
Telex: (312) 418 29 46

UNITED KINGDOM – ROYAUME-UNI
HMSO
Gen. enquiries Tel. (171) 873 8496
Postal orders only:
P.O. Box 276, London SW8 5DT
Personal Callers HMSO Bookshop
49 High Holborn, London WC1V 6HB
Telefax: (171) 873 8416
Branches at: Belfast, Birmingham, Bristol, Edinburgh, Manchester

UNITED STATES – ÉTATS-UNIS
OECD Publications and Information Center
2001 L Street N.W., Suite 650
Washington, D.C. 20036-4910 Tel. (202) 785.6323
Telefax: (202) 785.0350

VENEZUELA
Libreria del Este
Avda F. Miranda 52, Aptdo. 60337
Edificio Galipán
Caracas 106 Tel. 951.1705/951.2307/951.1297
Telegram: Libreste Caracas

Subscription to OECD periodicals may also be placed through main subscription agencies.

Les abonnements aux publications périodiques de l'OCDE peuvent être souscrits auprès des principales agences d'abonnement.

Orders and inquiries from countries where Distributors have not yet been appointed should be sent to: OECD Publications Service, 2 rue André-Pascal, 75775 Paris Cedex 16, France.

Les commandes provenant de pays où l'OCDE n'a pas encore désigné de distributeur peuvent être adressées à : OCDE, Service des Publications, 2, rue André-Pascal, 75775 Paris Cedex 16, France.

7-1995

OECD PUBLICATIONS, 2 rue André-Pascal, 75775 PARIS CEDEX 16
PRINTED IN FRANCE
(66 95 20 1) ISBN 92-64-14591-5 - No. 48235 1995